Your Home Fallout Shelter:

How To Ensure Your Family's Health & Survival In A Nuclear Incident

By Charles S. Brocato (Dr. "B")

Cover Photo: @jdoms-stock.adobe.com

All Other Photos: Charles S. Brocato

Dedicated to those brave Firemen & Plant Workers

Of Chernobyl

Who died because of "Lack of Truth"

Acknowledgements

Many thanks to my editor, Kathryn E. King, who aided me with every aspect of this book. Without her encouragement and help, this project would never have been completed.

Table of Contents

Introduction

Charles S. Brocato, known to his students as Dr. "B," covered radiation and why, in these interesting times we live in, you need a radiological meter in the first volume of Dr. "B"s Radiation Series titled *How To Choose A Civil Defense Radiological Instrument: Geiger Counters & Dosimeters.*

If you read this book, you have armed yourself in advance with at least one radiation meter and hopefully a dosimeter charger and several dosimeters. If you have these, then your first line of defense is complete in case of an atomic attack against the United States.

Your second line of defense will be your home fallout shelter, where you and your family can take refuge if a nuclear attack occurs and fallout rains down on unprotected America. When the Civil Defense program evaporated, so did all those well-stocked public fallout shelters. If an attack happens now, *you're on your own.* If you have no plan in place, you and your loved ones will suffer along with everyone else.

Much literature is out there about fallout shelters—from yesteryear. The government, back when it used to care about its citizens, funded many studies designed to learn the limits of overcrowding of shelters, *the limits of heat and humidity in public shelters, and the supplies a public shelter should stock to help bring the citizens safely through an atomic attack* so they can begin to rebuild the country upon leaving the shelter.

In this book, Dr. "B" examines topics such as what fallout is, why it is harmful to the human body, and how your home shelter should be set up to protect you from fallout. He also shows you how to stock it quickly and for relatively little expense, and how to know if a "worsening crisis" develops.

He has removed the technical jargon and physics that many people find incomprehensible and has made this book as interesting and as non-technical as possible, so that people can see immediately what they can do and hopefully get it done before anything happens.

If the worst happens and you need to actually use your shelter, Dr. "B" shows how to operate your home shelter in order to protect your family from the harmful effects of radiation and get safely through the crisis.

He has written this book without deep math and physics formulas, cutting straight to the heart of the matter: "staying alive and safe" during a nuclear incident!

Dr. "B"s mission in life is to save souls and save lives, and this Radiation Series is part of that mission.

1: *Why You Might Need A Home Shelter*

The idea of a home fallout shelter these days is gone right along with the Soviet Union, buried in the dim memories of the Baby Boomers who practiced diving beneath their desks in elementary school and grew up expecting a nuclear attack at almost any day.

During the 1950s and 1960s, if you were building a new home, you gave serious consideration to the idea of having a special room built in the basement or elsewhere on your property for that purpose. Or if you had a suitable spot in your yard, you could have a big hole dug and a prebuilt underground shelter placed there, and maybe you'd even receive a Bendix Family Radiation Measurement Kit as a reward for your purchase.

But never mind if you didn't have the means to construct your own family shelter—downtown in the basement of the bank and maybe several other massive buildings in the community, your local Civil Defense was prepared to set up public fallout shelters where you and your family could take refuge for as long as it took the fallout to decay.

In these days of "mutually assured destruction," many would find the very idea of building a fallout shelter laughable. Why, they wonder, do you want to be the only person left alive on earth, only to die of starvation in the "nuclear winter" or from the lingering radiation poisoning the entire earth?

As we explained in our first book, *How to Choose A Civil Defense Radiological Instrument: Dosimeters & Geiger Counters*, the ideas of mutually assured destruction, nuclear

9

winter, and long-lasting clouds of radiation are all myths, most likely designed to let the government spend all the money formerly allocated to Civil Defense on buying votes. As Congress passed more and more legislation that created and expanded the welfare state, they had to fund it somehow.

Voters, as always, are more interested in their own share of government benefits than in the possibility of a future nuclear war. More Medicare spending got a Congressman far more votes than a bill expanding the number of public fallout shelters, so Congress naturally grabbed onto any idea that showed Civil Defense was useless in the case of an atomic attack.

In the late 1960s and early 1970s, we stopped seeing signs on certain public buildings that designated them as Public Fallout Shelters. By the end of the 1970s, the idea of a fallout shelter had vanished into the graveyard reserved for outmoded ideas.

Civil Defense Replaced By FEMA

If something happens tomorrow, such as a natural disaster, we are assured that FEMA, the government entity that replaced Civil Defense in the late 1970s, will rush to the aid of those affected. The Federal Emergency Management Agency, however, in spite of its mission to lead in preparedness and response to emergencies, appears to have no mission to prepare Americans for a nuclear war.

Yes, they have "first responders" who are trained to operate radiation meters and secure areas of contamination. They also operate shelters for people affected by natural disasters, but in all our studies with FEMA, we know of no FEMA plans to operate an old-fashioned fallout shelter in case of a nuclear attack. The idea is most likely to set up a shelter outside the fallout zone, but the shelter would most definitely not be any protection should the wind suddenly change direction and deposit the fallout on the shelter area.

In short, with the demise of Civil Defense, the idea of public fallout shelters vanished. After all, what's the good of a fallout shelter if you're going to die anyway in a nuclear war?

As we explained in the first book, life always finds a way, and your job is to survive. If Jesus Christ overcame the last enemy, death, then death is the enemy of God, and our enemy also. We are to fight death and live for the glory of God, until the day He calls us home ... and not one day before.

If there is a nuclear war, assuming you are not in the Ground Zero area that is instantly destroyed, you will most likely survive, as did many citizens in the Japanese cities of Hiroshima and Nagasaki. When the bombs fell on Japan, nobody had ever seen an atomic bomb explosion before. Nobody knew what it was or what to do, and because of that, many suffered grave injuries that could have been avoided.

http://www.chemicalbiological.net/Hiroshima.wmv

Because these two atomic bombs were exploded in the air, fallout wasn't a huge concern for the Japanese people, but if they had been ground bursts, you can bet that fallout injuries would have been enormous because nobody knew anything about fallout, either.

During the 1950s and 1960s, many atomic bomb tests were carried out that gave scientists a lot of information about fallout, fallout patterns, and atomic bomb effects in general. This information was generally made available to the public during the Cold War era, but the moment the mutually-assured-destruction idea took hold, people saw no reason to prepare, and when the Soviet empire went belly-up, the whole idea of preparation for nuclear war went with it.

Now that relations between our nation and various other nations have deteriorated, it is becoming more and more clear to those who pay attention that nuclear war is once more on our horizon. In fact, it seems to us as if the United States is doing its dead-level best to provoke a war, whether to justify more expenditures on war machines or to distract the public from the real state of the U.S. economy, we don't know.

But we do know that if our government succeeds in provoking a nuclear attack, there aren't any more public fallout shelters out there for us any longer. So what are we supposed to do, especially if (as we expect) FEMA can't set up a shelter anywhere because of widespread fallout.

FEMA is a reactive agency, not a proactive one. As anyone who has been through a major hurricane or other disaster can attest, FEMA waits outside the area, and only comes in once it is safe to do so. In a hurricane, that means you won't see a FEMA truck until a good 3 to 5 days after the hurricane. If you hadn't made any advance preparations, you'd be mighty hungry and thirsty by the time FEMA got to your neighborhood.

If you can't count on FEMA to come rescue you with hot meals and trucks laden with MREs and bottled water, then what are you going to do in a worst-case scenario like an atomic bomb attack? Because we can tell you, FEMA personnel will be hiding under their beds just like every other unprepared American if this should happen. Or running frantically downtown in search of that old marble bank building that used to house a public fallout shelter.

The answer is simple, and it isn't one any of us particularly likes to hear, but we are going to have to be our own "first responders" in a case like this. By thinking about your response in advance, you and your family will have the best chance of survival.

If the United States should become involved in a nuclear war, your best chance of survival in good shape will come in the form of your own family fallout shelter, stocked in advance with simple preparations, just in case.

Why You Don't Want To Go To A Public Shelter

Civil Defense as an idea is dead in the United States, probably because of the expanding of the welfare state. People expect the government to do everything for them, and as we saw when Hurricane Katrina hit New Orleans, they also expect the government to give them hot meals. They didn't even want to read the directions on how to heat the MREs themselves. The government should have done it for them.

This is what you would be up against if the government did set up some public fallout shelters. Your family would share living space with people who are not the reserved, quiet, self-reliant people of the 1950s. Many people now are addicted to drugs of various kinds, both legal and illegal. They are also addicted to chips, dips, and Big Macs, and if they don't get them, they'll be quite hostile.

They are also addicted to noise, and you can bet that if any boom-boxes have survived the nukes, they will be found at the public shelters. Even if nothing electronic survives the EMPs created by the nuclear bomb explosions, some people are simply loud. Others will be having hysterics, anxiety attacks, and psychological problems of all kinds. Still others will see the shelter as one big storehouse to plunder.

Staying in a public shelter these days, no matter how gifted and charismatic the shelter manager may be, will be sheer chaos. People once knew how to follow orders when the situation called for it. Many people now, particularly those of the welfare underclass, think it is skin off their backs if they obey someone else's orders, and they appear to have no capacity to evaluate a situation and choose the course of action in their own best interests.

Since there are no public shelters in place now, that means that if FEMA managed to set up some, they have no food or water stored, no radiological instruments in place, and no trained shelter managers or radiological monitors. FEMA would have to depend upon receiving daily deliveries of food and water, and if there is an all-out nuclear war involving the United States, we can tell you that there will be no deliveries, no trucks running, and very likely no running water or electricity.

A Sign We May Never See Again

Nor will there be many FEMA personnel, no matter how public-minded and dedicated, who would be willing to leave their own families in such conditions to come manage a shelter, even in their own cities.

You Need Your Own Shelter

In such an eventuality, you must have your own shelter set up and ready to go. You can store water, the food your family eats, and you can have your own radiological meter. You can be in control.

Moreover, once you have chosen the space for your home shelter, you will have preparations that are ready to be used in the event of almost any disaster. Many areas have come under "shelter-in-place" orders when a nearby chemical plant experiences some unforeseen event, or a nuclear power plant within a certain distance of the area undergoes an unexpected event.

In the event of a chemical or biological attack, or if a large scale pandemic erupts, your home fallout shelter will serve you well. You will not have to join the rest of your community in a quarantine shelter.

Add to this the situation that occurred in a Boston neighborhood after the Boston Marathon bombing, when police literally shut down the neighborhood and forced citizens to remain in their homes while they searched for the perpetrators. We look for this sort of thing to be on the increase as the police become more and more militarized.

Creating your own home shelter is one of the best steps you can take to *stay in control of your own life, your own home, and your own family.*

So let us get started on what makes up a good home fallout shelter, first by understanding the nature of fallout.

2: *What Is Fallout?*

Fallout is exactly what it sounds like—something that falls out of the sky—after a nuclear bomb explodes. When a nuclear bomb explodes, Uranium-235 and Plutonium-239 atoms are split, or undergo *fission*, and the new atoms that are produced from this fission process are often radioactive themselves. Some are short-lived, existing only a few hours or days before they *decay* by spontaneously emitting particles and gamma rays from their nuclei until they become stable, non-radioactive atoms. Other fission products remain radioactive for years.

When an atomic bomb explodes at or near the surface of the ground, it blasts out an enormous crater, and all that dirt is sucked upward into the atmosphere along with the remains of the nuclear material and bits of the bomb itself. The vaporized and very tiny particles left of the bomb are full of highly radioactive fission products and they coat or cling to the particles of dust and dirt from the ground as the debris roils higher into the sky. Soon after the explosion, the bomb debris and dirt begin to fall out of the sky back to earth.

As you would expect, heavy particles of dirt and debris begin to fall back to earth immediately, so that the area immediately around the explosion crater receives much of the heavier fallout material.

The finer material can go quite high into the atmosphere, so high that the winds at those high levels are often traveling in very different directions than the winds near the ground. Since wind is the major factor that determines which way fallout travels, that upper level wind direction is usually the direction most fallout that concerns us will travel.

One thing in our favor is that the strength of the radiation given off by fallout tends to lessen as the radioactive isotopes contained in it decay. The longer the time the fallout remains aloft, the weaker its radiation is when it finally returns to earth. Radioactive decay is our friend, because it means that we don't have to stay in our shelters forever. Assuming no other atomic explosions take place, radioactive fallout usually weakens to such low levels of radioactivity that we can safely leave our shelters after a couple of weeks.

The Seven-Ten Rule

Scientists have worked out a method of estimating how long it will take radioactive fallout to decay to safe levels called the *Seven-Ten Rule*. It is, they state, not to be relied upon over the readings you take with your survey meter, and provides only an estimate. If another bomb should go off a day or two later, or if several bombs were exploded over several days' time, it cannot be relied upon at all.

This rule of thumb is based on the fact that fallout is most radioactive immediately after the explosion. After that, it tends to decay rapidly. Because of this fact, *the rule is only good for fallout of the same age.* Another bomb that explodes later and adds new, young fallout to the mix negates the rule.

The Seven-Ten Rule says that fallout will decay to one-tenth of its current strength in seven times the time you took the first measurement.

In forty-nine times the time you took the first measurement (7 x 7), the fallout will have decayed to one one-hundredth (1/10 x 1/10) the strength of your original reading.

18

You can use days, hours, or minutes for these calculations, but most people use hours for purposes of estimating when they may be able to leave their shelters.

You can view a video demonstrating the Seven-Ten Rule at:

http://www.chemicalbiological.net/Radiological%2 0Meter%20Page!.html#7-10 Rule

Predicting Where Fallout Will Come Down

Fallout prediction has become a fairly fine science, with lots of equations and predictive variables worked out, but as the researchers involved will tell you, it is far from a tried-and-true science. Think of it more as a forecast from a television weatherman: Lots of times he will be pretty much right, but lots more times he will be outright wrong because something unexpected happens that causes a whole new weather pattern to develop.

Some of the variables involved in these predictions include the kind of bomb and the yield of the atomic bomb exploded, the altitude of the explosion, and the meteorological conditions prevailing in the area when the explosion takes place.

Thus, fallout deposition depends upon and behaves a lot like weather. Since its direction depends on many things, including wind speed and wind direction, any little variation in one of those items will cause a change in the direction of travel and the area of deposition.

In general, we can think of fallout as "dust" that is composed of small and large particles thrown high into the sky. Where that dust would drop back to earth, so will the fallout from a nuclear explosion, with the larger, heavier particles dropping out first. If your house lies downwind of an important city or other strategic target likely to be targeted by an enemy bomb, you can usually assume your home needs a fallout shelter.

Most of the upper level winds over the American mainland are "westerlies," meaning they blow from the west toward the east. Thus, fallout over much of America would tend to travel toward the east.

One major study of the probable results of a nuclear attack on America involving some 250 nuclear weapons spread over strategic targets located all over the mainland showed that within 6 hours of the attacks, fallout would have spread to cover some 40% of the country. Within one day, fallout would cover close to 70% of the land. However, the study also pointed out that the greater the spread, the less strong the radiation given off, and the longer the time until deposition, the more the radioactivity of the fallout decayed.

Local Fallout Versus Worldwide Fallout

We are mainly concerned with "early" or "local" fallout, the particles that tend to return to earth within ten to twenty hours after the blast. Depending upon the speed and strength of the winds aloft, fallout can travel several hundred miles from its place of origin.

The larger, heavier particles usually drop out close to the point of the explosion. The smaller particles, carried high into the sky by the forces of the atomic blast, can be carried several hundred miles by high-level winds. Still other particles, even smaller, are carried into the middle level of the atmosphere called the troposphere. Most of these particles are deposited within a period of weeks along the latitude of the bomb explosion and can literally encircle the globe in four to six weeks.

Still other very small particles that are lifted to the upper level of the atmosphere, the stratosphere, can literally be stored in the stratosphere, from whence they travel around the world, with only a small portion falling out at any given time.

Fortunately, the *worldwide* fallout from the troposphere and the stratosphere usually is not dangerously radioactive. It tends to return to earth via precipitation. That in the troposphere is usually all deposited back on earth within a few weeks. That in the stratosphere, however, tends to remain aloft for months to years.

To learn more about fallout, go here:

http://www.chemicalbiological.net/Fallout.wmv

Nuclear Reactor Fallout

A nuclear reactor explosion can also result in radioactive particles being blasted high into the atmosphere if the explosion is powerful enough. At Chernobyl, the explosion was powerful enough to send out a cloud of radioactive particles that are still rendering the land in the direction of the prevailing winds in Eastern Europe and parts of Belarus and Russia uninhabitable after 30 years. Bear this in mind if you live anywhere near a nuclear reactor, especially downwind of it.

Emissions of radioactive gases from nuclear plants tend to affect mainly the areas near the plant in a downwind direction. If you live within a couple of hundred miles of a nuclear power plant, check on the direction of the prevailing winds in your area. This is a simple, self-defense precaution anyone can take, and the knowledge may help save your life.

How Does Radioactivity Harm The Human Body?

Fallout harms the body because it is radioactive, meaning that the radioactive components of fallout that cling to the dirt and bomb fragments constantly emit particles and rays of ionizing radiation.

Ionizing radiation is called *ionizing* because, when it strikes other molecules or atoms, it knocks electrons off the atoms and leaves them with an electrical charge (ion). When this happens, the atom or molecule is damaged, and if the atom or molecule happens to be a crucial constituent of a bodily cell, that cell is damaged and may die or be rendered inoperable.

The particles emitted, protons, electrons, and neutrons also cause damage when they strike molecules and atoms, particularly as they pass through the human body. If only a few pass through and damage other atoms or molecules, the body can usually repair the damage with no harm done. But if many particles and rays strike the body within a short time period, the body's repair systems are overwhelmed and cannot repair all the damage.

As scientists learned more about radiation and its effects on the body, they learned that certain tissues of the body were more susceptible to radiation damage than others. For instance, bone marrow, lymph system tissue, the gonads, the spleen, and gastrointestinal tissue are the tissues most sensitive to radiation damage. On the other hand, muscle and nerve tissues and the bone tissues of adults seem to be the least sensitive.

Researchers note that *acute* radiation doses are doses received in a short period of time that are high enough to cause damage to the human body, whereas a *chronic* dose may total up to the same amount, but it is spread over a period of time so that the body is able to repair damage.

Thus, *Acute Radiation Syndrome* results from a fairly high dose of radiation received within a short period of time. The earliest symptom is nausea and vomiting that is accompanied by fatigue, a feeling of discomfort (malaise), and a loss of appetite. These symptoms develop within 1 to 3 hours of the exposure. Blood changes occur in the form of destruction of white blood cells (part of the immune system) and blood platelets (blood clotting constituents). Within a couple of weeks, epilation (hair loss) occurs.

If the dose was not extremely heavy, the body can recover from Acute Radiation Syndrome, provided it has the proper medical support and no other conditions or injuries. However, for acute radiation doses above 600 Roentgens, complete recovery becomes more problematic.

Here are some dose limits that have been worked out by researchers over the years of studying the Japanese bombing victims and people who suffered high exposures in research and industrial accidents.

0 - 100 Roentgens: No vomiting, no organ damage, and no therapy required other than reassurance.

100 - 200 Roentgens: Moderate leucopenia (loss of white blood cells), up to half the people exposed will experience nausea and vomiting in 3 hours. However, the prognosis for complete recovery is excellent.

200 – 600 Roentgens: Above 300-Roenten exposures, all victims will experience vomiting within 2 hours. They will also experience severe leucopenia, purpura (purple spots on the skin), hemorrhage and infection. Epilation (hair loss) will occur in 2 weeks in those exposed to more than 300 Roengens. Treatment consists of blood transfusions and antibiotics, and deaths of 0 to 80% of those exposed will occur within 2 months. The Prognosis for recovery is good, but recovery may take up to a year.

600 – 1000 Roentgens: All exposed people in this range will experience vomiting within an hour. Victims will experience severe leucopenia, purpura (purple spots on the skin), hemorrhage and infection, and Epilation (hair loss). The prognosis for recovery is guarded at these levels. These exposures may require bone marrow transplants. Deaths of 80% to 100% of those exposed will occur within 2 months, with death due to hemorrhage and infection.

Above 1000 Roentgens: This is considered the lethal range, and the prognosis for recovery is considered "Hopeless." Vomiting begins within 30 minutes of exposure. Treatment is mostly palliative, and for exposures between 1,000 and 5,000 Roentgens, symptoms are mostly gastrointestinal, with death occurring within 2 weeks from circulatory collapse. For exposures above 5,000 Roentgens, deaths occur within 2 days due to respiratory failure and brain swelling.

Your aim is to keep yours and your family's exposure to any radiation as low as possible.

What Are You To Do?

Given that fallout can reach your home from so many different directions, what are you supposed to do if something outside your control happens, and your family is threatened?

You are going to place between yourself and the source of radiation the three things known to prevent radiation injury to people: *Time, Distance*, and *Shielding*.

3: *Where To Locate Your Home Shelter*

A fallout shelter is a structure designed to give you protection from fallout by supplying *shielding*, namely lots of dense material between you and the fallout.

If you set up your home fallout shelter in an area of your home as far distant from the source of radiation, the fallout, you have then employed the concept of *distance*.

Most people have no clue which areas of their home might provide the most shielding from fallout radiation. They think that if the home was not built with a safe room or a shelter, then there is no place that will be safe under that roof.

In fact, just being inside your home rather than outside when fallout is coming down or is already present cuts the radiation dose you receive by about half. This is because the roof and walls of the house provide some shielding from fallout radiation.

Two of the old Civil Defense publications that give good information about building and stocking fallout shelters can be found here:

http://www.chemicalbiological.net/In%20Time%20 of%20Emergency.pdf

http://www.chemicalbiological.net/guidance%20for %20fallout%20shelter%20stocking.pdf

Basement Shelters

If your home has a basement, this is usually one of the best places to locate a home shelter. If it is an underground basement or even partially underground, so much the better. All that dirt surrounding the sides of the basement gives some of the best protection available (lots of mass between you and the radiation), and the basement is usually as far from the radiation as possible (distance),

If your home has a basement dug into the ground or rock, this location may be ideal. Find the basement wall that has no windows or openings to the outside and build your shelter here. You can construct a small room within the basement by stacking totes of books or positioning file cabinets or other solid pieces of furniture as walls.

On top, place an old door or a sheet of plywood and pile the plywood high with totes of old clothes, books, or other items that will provide a lot of mass (shielding) for radiation to plow through before it reaches you.

If you have a sturdy, large desk or table in the basement, or even a counter, these can serve as the basis of your fallout shelter. Use the top of the desk, table, or counter as the roof and position heavy furniture, boxes of goods, file cabinets and so forth around it and on top. The more mass you can place between you and the radiation, the better you will be shielded.

You can also build a low-cost shelter within your basement by using solid concrete blocks, a known form of excellent shielding against radiation. During the 1950s, Civil Defense publications featured various plans for building a shelter within your basement, using bricks or solid concrete blocks.

If you have a basement large enough for this, you can find plans by downloading "Family Shelter Designs" at:

http://www.chemicalbiological.net/FamilyShelterD esigns.pdf

Cresson Kearny's book, *Nuclear War Survival Skills*, also contains plans for "expedient" basement shelters, meaning a shelter you can put together fairly fast if you should receive word that a nuclear attack is coming. You can order a print copy of his book from Amazon.com, or download a free PDF copy from:

http://www.chemicalbiological.net/nuclear%20war %20survival%20skills.pdf

Outdoor Shelters

If your backyard has plenty of room, and you are located in an area where the water table is ten feet down or deeper, you may want to look into a trench shelter or other earth-covered shelters. Cresson Kearny, in his book *Nuclear War Survival Skills*, gives many simple plans for such shelters. All you need are shovels and a suitable area to dig. The Civil Defense publication, "Family Shelter Designs," also gives some out-door shelter plans.

But what if you don't have a suitable basement or backyard, and your only option is somewhere inside your own home or apartment? In that case, here is what to do.

First: Find An Inner Room

Bearing in mind the two ideas of distance and shielding, the best area in your home for a shelter, assuming you have no basement, will be an inner room with no windows. This can be a walk-in closet or even an inner room where you can construct a simple room-within-a-room shelter out of furniture, totes of books or clothing, or bookshelves loaded with books. What you aim to do is add another layer of shielding between you and the radiation on all sides.

You should try this construction now so you can see if you have sufficient items to use. Bookshelves and file cabinets make suitable thick walls. Old doors or a large sheet of ½ inch or thicker plywood can serve as a roof. Make sure the 'roof' is strong enough to hold another layer of book-filled totes to shield you from radiation coming from fallout deposited on the roof of your house.

If you have a closet that is big enough, line the outer walls around the closet with your file cabinets, totes of books, bookcases, or other pieces of furniture that will serve as another barrier to radiation. In this case, the thicker your fortifications, the better. Then get into the attic above the closet and place another layer of bulky items, or even bags of soil, to further shield you from radiation coming in through the ceiling.

Inside the shelter area, you should have room to place chairs or lots of pillows so that your family members will have comfortable places to sit or lie down during the time they will need to spend in the shelter. If you have a sofa that will fit inside, that would add to your comfort.

Second: Shield Your Shelter Space

Remember, the idea of a shelter area is to shield you from receiving a large dose of radiation. Anything you can do to thicken the walls or ceiling will add more protection.

Scientists have discovered that the best shielding from gamma radiation consists of materials that contain elements with a lot of electrons in their outer shells. The more electrons present in the outer shells of atoms, the more likely the incoming gamma rays are to run into an electron and be weakened or deflected.

Earth, water and concrete contain many electrons that will deflect incoming radiation and are readily available to you. The best shielding material is lead, which is a heavy, dense metal loaded with outer-shell electrons. Since lead isn't something most households have lying around, protection ability of various materials against gamma radiation is usually given in relation to so much concrete.

According to *Protection in the Nuclear Age*, in terms of shielding capabilities, 4 inches of concrete has the same shielding capacity as:

  5 to 6 inches of brick

  6 inches of sand or gravel

  7 inches of earth

  8 inches of hollow concrete block

  10 inches of water

  14 inches of books or magazines

  18 inches of wood

However, lead is available to those who think ahead. You can buy lead shielding at plumbing supply stores where it is sold as sheets about ¼ inch thick, for plumbers to construct vent pipe flashing for home plumbing. You can buy several sheets of this relatively inexpensive material and use it to block holes or windows, or lay on top of the roof of your shelter as another layer of shielding.

How Thick Should Your Shielding Be?

According to *Radiation Safety in Shelters*, concrete is the material all other materials are rated against in terms of how thick a shield made of it needs to be. That 4 inches of concrete mentioned earlier, if used to build walls for a shelter, actually needs to be about 3 feet thick to absorb most of the gamma radiation from fallout striking them. Lead, the best shield of all, need only be 0.2 times as thick to provide as much shielding as concrete. So 3 feet of concrete would equal a 0.6 foot (or about 7.2 inches) thick sheet of lead.

Of the things readily available to you, a homeowner or an apartment dweller with an inside shelter, earth needs to be 1.4 times as thick as concrete to be equal in shielding value. Bricks made of clay should be 1.4 times as thick. Glass needs to be 0.9 as thick, and hardwood like maple or oak should be 3.3 times as thick. Plywood needs to be 5 times as thick, wallboard 2.7 times as thick, and steel 0.3 times as thick.

As for your totes full of items, slick magazines need to be 2.5 times as thick, newspaper needs to be 4 times as thick, and water should be 2.3 times as thick as concrete. Clothing wasn't given, but you can bet that totes full of stored clothing, or simply stacks of clothing from your drawers or closet will probably need to be stacked about 3 times as thick as 3 feet of concrete. Make use of what you have, and stack it as high or as thick as you can.

The human body, as a matter of interest, needs to be 2.5 times as thick as concrete to provide the same shielding. This may become important if you have small children or a pregnant woman in your shelter who will need special protection from radiation. *The adults present can encircle the more vulnerable occupants and provide extra shielding with their bodies.*

Soil And Water As Shielding

You can use bags of soil bought from a nursery, or make up your own bags of soil using pillowcases or plastic bags. You can also place your water storage barrels around your shelter, and store buckets or plastic 5-gallon containers full of water on the roof of your shelter. A plastic toddler's swimming pool, available during the spring and summer at most "dollar" stores, is excellent for this purpose. Cover open water containers with a tarp or plastic shielding, as the water stored here can serve as part of your drinking water storage.

Remember, we are interested in protection from radiation, and in a pinch, anything you can pile on the roof of your shelter without causing it to collapse will serve the purpose.

If you own a particularly sturdy table, a dining table or any other kind, you can use this table as the roof of your shelter and place heavy furniture and totes of books, etc. around it, and store more water and totes on top of it.

Design & Draw Your Shelter

Once you have planned how to put your shelter together, remember that the door should have a heavy file cabinet or bookcase placed about a foot or two in front of it, such that you must slide in sideways to get into your shelter area. This creates a baffle that will stop radiation from coming directly into your shelter via the entryway.

31

Remember, radiation travels in straight lines, and a baffle will stop a lot of radiation rays from coming directly into your shelter via the entrance.

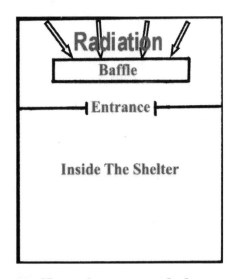

Place A Baffle Before Your Shelter Entrance

The shelter needs enough room inside for you and your family to sit comfortably and move around a little. Naturally, the more room you have, the better, but in a small inner room or closet, or in an apartment, all the space you'd like may not be an option. In this case, you must opt for the best space possible to block incoming gamma radiation. If you have to sit around for a week or two, so be it. When you're with your family, the extra closeness is easier to take.

Once you have achieved a good placement of your shielding furniture and other items, you should make a drawing of your planned shelter area, with dimensions noted, along with placement of heavy pieces of furniture, and a list of what you intend to construct it from. If you have to put the shelter together in a big hurry, these detailed plans will help you remember where you had planned to put which piece of furniture.

Once you have a good drawing made, scan it or go to a copy shop and have copies made. Once fallout begins, these copies are useful to record daily radiation readings in each area of the shelter. This information may be invaluable to you after the crisis and will help you figure up the doses of radiation your family received, not to mention tell you which areas of your shelter may need shoring up if you should need it again.

When the time comes to actually put your shelter together, and you have moved all the heavy pieces of furniture into place, you will no doubt notice many cracks, crevices, and outright holes that radiation could come through unimped-ed. You can stuff the holes or crevices with clothing, or you can bend the malleable lead sheathing from the plumbing store around cracks or holes.

Basically, your aim is to do anything you can to make radiation travel through more material before it reaches you and your family members.

Dr. "B"s Under-The-House Shelter Plan

Some people still have houses on piers, which usually means there is a good crawl space beneath the house. If you have sufficient space under the house (and some houses have almost enough space beneath them to stand up), you can enclose an area to serve as a good fallout shelter. Your family should at least be able to sit up if you decide to create your shelter here, but in a worst case scenario, nobody will mind spending most of their time in the shelter reclining.

One plus to this choice during the heat of the summer is that the area beneath a house often tends to be cooler than inside the house. If it is relatively dry beneath your house (slight dampness doesn't matter) and standing water doesn't tend to develop in rains, this might be a good choice for your home shelter.

Here is a method of building a shelter beneath your dwelling that you can do relatively easily if you have time and the garden shops or hardware stores that sell bags of soil or sand are still in business.

Beneath your house, mark off the area you intend to use as your shelter. It can be as large or as small as you wish, and you can even create "rooms" if you desire. The entrance should be located at the edge of your house, with your shelter area taking up as much or as little of your under-the-house space as you like.

Begin buying bags of soil or sand, ten or twenty at a time, depending upon the size of your vehicle. These bags generally weigh about 40 pounds each. We advise that you buy the bags free of wood chips, such as sand or plain soil rather than "top soil."

These bags of soil are the *building blocks* for the walls of your under-the-house shelter. You are going to stack the bags, one on top of the other, up to the floor of your home. The floor of your home will act as the shelter roof.

Drag the bags beneath your house using an old blanket, one or two bags at a time, and begin stacking them, one upon another, to build your walls. This may take a while if you're hauling bags of dirt from the garden store, but if you've got the time, you can build quite a nice shelter out of bags of soil. Using just one stacking of bags, depending upon whether you turn them lengthwise or crosswise, you will have a dirt thickness of about 1 foot or 2 feet, respectively.

Remember, an ideal wall of dirt for radiation shielding should be about 4.2 feet thick, so you may want to make a doubled wall of sacks, at least two bags thick, maybe even three.

About three feet in front of the entrance to your shelter, which should be at the edge of your house, build a wall that extends two feet beyond each side of the entrance to serve as a baffle. This is to keep radiation from entering directly through your entrance while leaving your entrance open for ventilation. You can also build another short wall just inside the entrance to the shelter to serve as a second baffle.

Remember, the house above your head provides some shielding, but you should increase your shielding by putting at least one layer of bags on the floor inside your house to cover the area above your shelter. You may want to place plastic sheeting or a tarp on the floor first to protect your floor and lessen the amount of cleaning to be done later.

Using Sandbags To Build Your Shelter

Using Books Above Your Shelter

If the ground beneath your house tends to remain damp, spread plastic sheeting or a tarp over the ground that forms your floor, then spread blankets, sleeping bags, pillows, bed rests, and whatever else will increase your family's comfort.

If you have the time and inclination and available space, not to mention the bags of dirt, you can partition off rooms in this shelter. If you want to partition off a space and do not wish to build walls, you can use a sheet, blanket, large piece of cardboard, or a suitable piece of plywood to partition off a room. At least one small portioned-off area is usually needed to provide privacy as a toilet facility.

In the toilet area, dig a hole in which to place your portable toilet. Dig down into the earth about half to two-thirds the height of your portable toilet, with a circumference about 2 inches wider than the toilet. Drop in your portable toilet and seat it in the earth hole, tamping dirt around the edges to hold it steady.

Place the seat on the toilet, and line the toilet with two white plastic garbage sacks. Add enzymes and disinfectants.

Remember to stock plenty of toilet paper, plastic garbage bags, enzymes and disinfectants in your under-the-house shelter for toilet use and sanitation. Also have hand sanitizer and paper towels to maintain personal cleanliness.

Again, make a drawing of your shelter area and have copies made so that you can keep daily records of radiation readings in the various areas of your shelter.

When you start storing bags of soil for the building of this shelter, don't forget to include enough bags to cover the "roof" area of the shelter, i.e. enough bags for you to place on the floor inside your house that lies above your shelter area. Although the house and floor provides a fair amount of shielding from radiation, you want all the extra protection those bags of soil can supply.

4: *What Civil Defense Once Had For You*

Back in the 1950s and 1960s, a "government shelter program" existed which called for local Civil Defense organizations to pick out certain public buildings with suitable areas that could be used as public shelters in cases of nuclear war for protection from fallout. The government stocked these shelters with certain basic supplies designed to keep up bare nutritional standards during the time of "shelter living."

Our Radiological Meter Page features a video which will give you a rundown of what the government used to supply for each person who might take shelter in one of these facilities.

http://www.chemicalbiological.net/Radiological%20Meter%20Page!.html#What The Government Used To Supply

Study this and weep, because these public shelters are long gone now. There will be nothing less than total chaos on the streets during this time.

Any fallout shelters set up now would be set up by FEMA after the bombs fall, and very likely after the immediate fallout has come down also—if they even come into an area where fallout has already been deposited. This is very doubtful, especially if several cities are hit at the same time or multiple bombs have fallen all over the United States.

However, back in the 1950s, the government put some of the best researchers to work in recommending what supplies would be needed for the public shelters, and we can learn a lot from what was once supplied for the shelters and why.

Federal Government Supplies For Public Shelters

Once the local Civil Defense organization set up a public shelter, the government had certain supply "kits" that would be shipped to the shelter for storage until they were needed. In the meantime, the local Civil Defense people were responsible for keeping a check on the supplies. If the local organization thought it necessary, they could add to the basic items supplied by the government.

Interestingly enough, the government supplies appeared to be geared to last about 5 to 11 days, although most of the available literature calls for a shelter stay of about 2 weeks in order to safely avoid the heaviest fallout radiation. For this reason, you will likely find the government's food rations fairly sparse.

Food

The government food rations consisted chiefly of special wheat biscuits, or crackers. The crackers were sealed into rectangular tins that were packed in fiberboard cartons for storage. Packed like this, these "survival crackers" would likely last for years.

Government plans indicated that the rations supplied, if the shelter held the number of people it was rated for, each person in the shelter would receive 10,000 calories for a 14-day stay. This would amount to 2,000 calories a day for 5 days or 900 calories per day for 11 days, and about 700 calories per day for 14 days.

The crackers consisted of "crackers, biscuits, and bulgur wafers." The crackers were supposedly similar to Graham crackers, but what the difference between the crackers and biscuits was, we aren't sure. The bulgur wafers were made from a cracked, parboiled wheat (bulgur) and may have been provided for bulk-laxative purposes.

The government also supplied a "carbohydrate supplement," which consisted of hard candy pieces in bright colors. The purpose of the carbohydrate supplement was to add more calories to the diet in a concentrated form.

We presume that if the shelter had more occupants than allowed for, each shelteree would likely suffer certain cuts in the amount of calories allotted per day. The same would likely happen if more fallout arrived and people had to remain in the shelter for longer than 11 days.

However, if some of the people brought food supplies with them, or if the local Civil Defense organization had added to the supplies, then the rations would last longer and supply a bit more variety. This would also work the other way if there were less occupants in the shelter than planned for. Each person could receive more calories per day, and perhaps more variety.

Water

The government supplied enough water storage drums, each holding 70 quarts of water, so that each occupant (if the shelter was at its rated capacity) would have 14 quarts of water per person. For a one-week stay in the shelter, that would give each person 2 quarts of water per day, with less if they stayed longer (1 quart per day for 14 days) and more if they stayed less than a week.

Sanitation

The government-supplied Sanitation Kits contained plastic cups, enough to allow for extra people in the shelter or for breakage of some people's cups. Each cup was apparently assigned to a person, and that cup was to last the person until the shelter stay ended.

The government also included a siphon hose to aid in transferring water from the water barrels to pitchers or individual cups.

A barrel packed with 20 rolls of toilet tissue, a toilet seat, and sanitary napkins was included in the Sanitation Kit for use as a commode.

For the disposal of human wastes, the barrel was supplied with a double plastic liner to prevent leakage, a tie wire to tie up the plastic bag when the barrel was full and plastic gloves to use in disposing of the filled drum.

The kit also included a waterless hand cleaner and a spatula to dispense it and a chemical disinfectant to help in controlling germs and odors.

When the water barrels were emptied, one by one, the empty barrels were then intended to serve as commodes by transferring the commode lid to the empty barrel.

First Aid & Medical

The government-supplied Medical Kit contained basic medications like aspirin, penicillin, and sulfadiazine; alcohol; tongue depressors; a thermometer; cotton-tipped applicators; cotton; and gauze pads, plus two booklets providing guidance in treating common illnesses and complaints.

Radiologic

The government also provided a Shelter Radiologic Kit that contained everything shelter radiation monitors would need to keep track of radiation levels in the shelter, outside the shelter, in food and water, and on people.

The CD V-700, a low-level radiation survey meter, also known as a Geiger counter, was included to do the job of measuring small amounts of radioactive contamination, as in foods or water, or on people.

The CD V-715, a high-level radiation survey meter, would be used to keep track of radiation levels inside and outside the shelter, so that the radiation monitors could move people to safer areas of the shelter if necessary.

The CD V-750 dosimeter charger and the CD V-742 dosimeter were included to keep track of the total radiation dose received over the shelter period.

This kit even contained batteries for the equipment. Indeed, with this selection of radiological instruments, the shelter radiation monitors had everything they needed to keep track of radiation levels and contamination, not to mention the total dose of radiation received by the shelter occupants during the period they remained in the shelter.

What if You Wanted To Stay Home?

As you can see, the government had thought of literally almost everything that would keep a person alive and functional in a public shelter over the two weeks or so they remained in the shelter.

But if you wanted to shelter at home with your family, the government tried to help you do so. It put out quite a slate of publications on how to build a home fallout shelter and how to stock the home fallout shelter.

The home-shelter occupant was advised to pay attention to the radio for advice on what the radiation levels outside were and when to leave the shelter. If the homeowner bought a prefabricated fallout shelter and had it installed, it often came with a free Bendix Family Radiation Measurement Kit, consisting of a dosimeter charger and two special dosimeters. One dosimeter measured the total radiation dose, all the way up to 600 Roentgens. The other was dubbed a "Ratemeter" and was designed to give an idea of the dose rate in Roentgens per hour, or how heavily the radiation was coming in.

What Happened To All Those Supplies?

In 1969, government stockpiling efforts for the shelter program ceased. Studies indicated that the cereal-based foods stored in the shelters (the crackers, biscuits, and bulgur wafers) had likely become rancid, so in 1976, the government sent out a circular to the local Civil Defense organizations giving "guidance" for disposing of shelter supplies.

A circular was also issued saying that the shelter medical kits contained supplies that were past the specified shelf-life of five years and ordered local governments to destroy the medicines in the kits, as they had probably deteriorated. But the sanitation kits were still good and should be kept.

Today, we don't have any guidance from the government, other than these old publications from the Office of Civil Defense. From our studies, these older publications told the truth about radiation and how to survive it, whereas *most of the newer information seems unable to arrive at a conclusion about what to do.* Yet, the laws of physics have not changed!

That's why we say, **imitate** the best government research of days gone by then improve your preparations with any modern advances that have come along in the last 50 years.

Although it seems to us as though government preparations for nuclear warfare have gone mostly backwards in the era of "mutually assured destruction," surely there are some improvements and some new knowledge we can take advantage of in stocking our home shelters.

5: *Imitating Government Supplies*

Government supplies for public fallout shelters were chosen for mostly good scientific reasons. That's why we can benefit by mimicking, in our own home shelters, the kinds of supplies stored in public shelters.

By setting up your shelter in your own home, you can also take advantage of whatever you normally have in your kitchen cabinets and pantry, plus your medicine chest and other home items. You'll be much more comfortable and eat a more varied diet than you would in a public shelter—and that's assuming there are any public fallout shelters set up if something should happen in the near future.

Your Survival Crackers

The government stockpiled its "wheat biscuits" or crackers for a very good reason. Under shelter conditions, you will be largely inactive for something like two weeks. You will move around very little, and water may be in short supply. Heating or cooking food may be out of the question. Heat and humidity can be extreme problems in a shelter, especially during the summer months, and you will want nothing that adds heat in any form.

 Scientists recommended a high-carbohydrate diet because fats and proteins require more water for digestion. So the cracker was a logical choice for that reason.

Another reason is that crackers are somewhat similar to a substance called "hard tack," which used to be used as the major food for sailors on long voyages and soldiers in the field. Hard tack was basically composed of flour and water that was baked at low heat until all moisture was removed. It would keep for months, so long as it was kept dry. The only problem came from the fact that the dryer and harder it was, the longer it kept, and it could get pretty hard. Sailors or soldiers would often use it as a plate to hold the rest of their food. If the food didn't help soften it, they had to dunk it in their coffee.

Thus, we have the perfect survival food in these special crackers. They are relatively high in carbohydrates and calories; they require less water for digestion; and they are a dry food that keeps very well for long periods of time.

We are not recommending saltine crackers as found in grocery stores. They contain too much salt and are usually made with various vegetable oils, which are not good for your health, especially under stressful conditions or conditions involving radiation. Plus they do not store well beyond a few months, thanks to those oils rapidly going rancid.

Instead, we recommend the old-fashioned "pilot crackers" or "sailor's bread," best bought from food storage companies like Mountain House or AlpineAire, already packed in a #10 can for long-term storage.

Interestingly, these crackers even fit the logistics that shelter management manuals describe for the public shelter food plans.

Mountain House is probably one of the best-known food storage companies, and it makes a "Pilot Cracker" that has 60 calories per cracker. For a shelter diet totaling 900 calories per day, that would amount to 5 crackers at breakfast, 5 crackers at lunch, and 5 crackers at supper. AlpineAire makes a very similar cracker, the Royal Kreem Cracker.

Mountain House Pilot Crackers

AlpineAire Royal Kreem Crackers

A Hawaiian company, Diamond Bakery, makes an original "Saloon Pilot Cracker" that is also made with a little suet. These crackers contain about 65 calories per cracker and would also serve well for shelter storage, but their packaging is such that you would need to repack them for long-term storage yourself. You may wish to try a package before deciding on these crackers over Mountain House or AlpineAire.

Hawaiian Saloon Pilot Crackers

The public-shelter diet was eked out by a "carbohydrate supplement," namely small pieces of hard candy. For this, you can choose any hard candy you like that is made with real sugar rather than high-fructose corn syrup. This is harder to find than you might think. Read the labels.

We have checked out some hard candies and discovered it is almost impossible to find a hard candy that does not contain corn syrup. When you go looking for hard candy for storage, look for sugar at the top of the list, and corn syrup further down. Avoid candy containing high-fructose corn syrup.

The public shelter diet could also be made more interesting by adding "spreads," such as peanut butter or jam. Peanut butter is not particularly storable for long periods, but if you tend to keep peanut butter on hand in your kitchen cabinets, simply keep 3 or 4 jars on hand all the time and rotate them. When you start a new jar, buy another.

We recommend one of the natural peanut butters made with palm oil, or a fully-hydrogenated peanut butter rather than a partially-hydrogenated because of the trans fat content of partially hydrogenated foods. Jif Peanut Butter is a good fully-hydrogenated brand. Avoid all peanut butters with high-fructose corn syrup added.

Generally speaking, one tablespoon of a good peanut butter contains about 100 calories, and it can definitely add "interest" to a diet composed chiefly of survival crackers.

Jams or jellies favored by your family can also be stored almost indefinitely. If the jar is not opened, jams and jellies will last for many years in great condition. But once more, you must read labels because many formerly good brands are now made with high-fructose corn syrup rather than real sugar. You want jams and jellies made with real sugar, which helps stop the stress response you will likely have if you and your family have taken shelter in your home to avoid fallout radiation.

A tablespoon of jam or jelly usually contains around 50 calories.

This high-carbohydrate "shelter diet" requires far less water for digestion than the higher fat and protein diet usual with most long-term storage and freeze-dried foods. This is why we recommend several cases of Pilot Crackers be stored especially for use if you should have to use your home fallout shelter. If heat buildup is a problem in your shelter, you will be glad to have no-cook food on hand.

But you will still need water, and at a time like this, you would be wise to plan on using your own stored water rather than take chances with water that may possibly be contaminated with fallout.

When you realize there is a worsening crisis in the news, that is a good time to check on your water storage. Many food storage companies also sell water storage barrels of various sizes. The 55-gallon barrels are the most popular size, as one barrel contains 220 quarts of water. If you have four family members in your shelter, that is enough to give each member 2 quarts of water per day for at least 25 days, with some left over for other uses. If you have several storage barrels, so much the better.

This would not be a time to plan on using your swimming pool's water for anything other than flushing the toilets. Even if your pool was well-covered, fallout particles would likely find a way into the water, probably every time you lift the cover to dip some water out.

The preparedness companies are now advertising a 320-gallon water storage tank that is almost 8 feet tall. If you have a place to put such a storage tank, it would probably be an excellent buy for a large family.

Also, if you know an attack has occurred, but fallout has not yet arrived, you can fill your bathtub and every available container you have on hand with water. Just make sure these containers are covered, because once fallout arrives, it will behave like dust. As dustbowl dwellers discovered during the times of dust storms, no matter how tightly they sealed up their homes, the dust always found its way inside. You don't want any fallout particles in your drinking water.

Radiological Instruments

If you should have to spend one or more weeks in your home shelter, you will want to know the radiation levels both in your shelter and in your home and the areas surrounding you. The government back in the old Civil Defense days told people to have a battery-operated radio in order to know when radiation levels had dropped to safe levels, but these days you would be wise to check out those levels for yourself.

It isn't that you don't trust the government, assuming it remains up and running and actually has a means of monitoring your particular area, it's just that you'd rather do it yourself and compare your results to theirs. After all, readings can vary widely even within the same area. Fallout has a way of settling all in one area and leaving other areas almost void of particles, much in the way snowdrifts form.

The only way to know this is to have a radiation meter and to know how to use it. For aid in choosing a radiation meter to serve your needs, see Vol. I of our Radiation Series, *How To Choose A Civil Defense Radiological Instrument: Geiger Counters & Dosimeters.*

The Four Most Popular CD Meters

CD V-700, CD V-715, CD V-720, CD V-717

If you can't afford a good quality, calibrated meter, then you should try and obtain one of the Bendix Family Radiation Measurement Kits from e-Bay and learn how to use it. One way or another, you need to be able to measure the radiation around your particular shelter and inside it.

Bendix Family Radiation Measurement Kit

First Aid Supplies

It goes without saying that if you must spend time in your home shelter, you may need a first aid kit. The first aid kit that you build should fit you and your family, in terms of your personal needs.

If you or anyone in your family takes prescription medications, it would be wise to have a stockpile of those particular medications. If a nuclear attack occurs in America, you really have no way of knowing when the pharmacies will be up and running again and able to dispense your prescription ... if they ever do get up and running again.

Above all, you want to prevent a "Katrina" scenario, where many of the people rushed to the shelter without their insulin or other required medicines. They believed that "someone," namely the government, would take care of the matter, and if not, why not? When nothing was forthcoming, because the city was flooded and the pharmacy owners had fled the area, they enacted a first-class melodrama, swooning and moaning.

The way to avoid being without needed medicine is to have enough on hand to sustain you. Granted, it is hard to get ahead on some medications because of government regulations, but generally you can get a month or two of pills stockpiled on most blood pressure and heart medicines.

You should always have some broad-spectrum antibiotics on hand. During your shelter stay, it is always possible that someone could develop an infected tooth or some other infection, and it will usually be during a time that you most do not want to leave the shelter to seek treatment. If you have supplies on hand, you can treat the person yourself until the fallout decays and you can leave the shelter and find (we hope!) a doctor.

Other supplies to have on hand are gauze, bandages, cotton-tipped applicators, oral thermometer (mercury if you can still find one), safety pins and regular Band-Aids in various sizes. You should also have items like aspirin, Pepto Bismol, Calamine lotion (for itching), antibiotic ointment and petroleum jelly, plus any other item your family normally uses. You should also have a box or two of vinyl gloves.

Also good to have on hand would be the *Merck Manual* (not the Home Edition) and a good first aid book such as the *ACEP First Aid Manual*. These are for consultation when you suspect someone in your shelter is ailing and you want advice on how to treat the person.

Sanitary Supplies

Sanitation is extremely important in a shelter, even if the shelter is in your home and you have access to your own home bathroom. In a nuclear war situation, water may not be running and toilets may not be flushing.

If you have a swimming pool, you can dip buckets of water from it for purposes of flushing your toilet. Most people don't know that if you pour water directly into the toilet bowl after using it, the toilet will flush on its own, without flushing. Do not use more water than necessary to cause the toilet to flush.

However, if you don't have ready access to a source of water for toilet flushing, this is when your portable or emergency toilet comes into use. Be sure you have plenty of plastic garbage bags for use in the toilet and boxes of vinyl gloves especially for sanitary usage.

When you set up your portable toilet for use, place one packet of toilet enzymes, available from various preparedness stores, in the bottom of the sack. These enzymes help to break down wastes and prevent odors. If you do not have any toilet enzyme packets, use Lysol concentrate. Pour a couple of tablespoons into the portable toilet after each use to keep down odors.

http://www.chemicalbiological.net/Keep%20It%20
Clean.wmv

It goes without saying that this is a really good time to have plenty of rolls of toilet tissue stored. Once a nuclear event occurs in America, chances are no trucks will be rolling and no stores will be open, so if you happen to be down to your last roll when it happens, you will likely be out of luck.

During your shelter time, if you have no running water, the best thing to do when the toilet bag is getting full is to bury it. You must bury it deep enough so that animals can't dig it up and flies can't find it.

One thing you must prevent during this time is insects such as flies, which are known disease carriers. Practicing strict sanitation is the best way to do this. See:

http://www.chemicalbiological.net/Fallout
Shelter_II.wmv

How Long Will You Remain In Your Shelter?

The one good thing about fallout is that it decays. Generally speaking, depending upon how much fallout your area receives, most fallout has decayed to safe levels within a week to two weeks.

The problem with this timeline arises when nuclear bombs are exploded at intervals, for instance, a second attack follows the first within a few days. This may mean a fresh delivery of fallout to your area, which means that the 7-10 Rule can't be used and you're back to Square One as far as leaving your shelter goes.

This is why you need a meter. Without a local Civil Defense station monitoring the fallout in your area, chances are you will not know what the levels are unless you can measure them yourself.

We think a good length of time to plan on remaining in your shelter is about two weeks. If your meter tells you the levels have fallen within a week, that's great. But if there should be follow-up attacks, all bets are off and you may need to remain in your shelter for several weeks.

Only your meter can answer that question when the time comes.

6: *What Else Is Needed For Your Shelter?*

You know you need plenty of toilet paper, white plastic kitchen garbage bags and other sanitary supplies for your shelter, but what else might be needed to make your life more pleasant? Two important things come to mind immediately: space allocation per person and ways to deal with heat and humidity.

Space Considerations

During the Civil Defense days, many studies were done to learn just exactly how much space each person required during a shelter stay for optimum well-being. Before that, the military, in particular, the Navy, investigated this same question, for obvious reasons. A perusal of the many Civil Defense papers done at that time tells you that anywhere from 9 square feet per person to 14 square feet per person is considered ideal.

For this reason, *Safety In Shelters*, one of the best Civil Defense publications on public shelter operations, settled on 10 square feet per person as the minimum space requirement. They also pointed out, as do most of the publications, that the more space per person, the easier the shelter stay will be on everyone there. When people are elbow-to-elbow, a shelter stay is anything but pleasant.

Obviously, if you do not have a dedicated fallout shelter and you must rely on an "expedient" shelter that you construct beneath a table or in a corner of your basement, you know your space allocation will be far less than this. All you really need is enough space to sit and lie comfortably for the duration of the shelter period.

But you are not in a public shelter, where you would strongly dislike being crowded up with strangers. You will be with your own family, and if conditions are more crowded than is comfortable, you will still be able to weather the time of confinement far better than you would in a public shelter.

What many of the publications pointed out also was that in times of need, the shelter could be far more crowded than it was designed for, and people could still make it through in good shape.

We agree that they can if they understand the necessity, but the problem today is that many Americans, especially the "entitlement" segment of the population, do not understand the necessity and they do not seem to have the ability to understand it. In their opinion, somebody ought to do something about it, and that someone certainly isn't them.

However, in your own shelter, you and your family, knowing the necessity and monitoring the radiation levels with your own meter, can weather just about anything.

There are only two things that are extremely difficult for anyone to weather, no matter how great his understanding. In fact, according to *Overcrowding Potential*, "it appears that physiological stress is more important in determining performance during the shelter period. And these [two] environmental parameters largely determine the level of physiological stress."

These two "environmental parameters" are **Heat** and **Humidity**.

Heat Considerations

Heat is a definite problem during the hot summer months, and if an attack occurs during the summer, heat can be a huge problem in your home shelter.

In a public shelter, with many bodies crowded into an area that may not have adequate ventilation and with no electricity, heat can be a real killer. Each human body gives off about 400 BTU/hr when at rest, and perhaps 300 to 320 BTU/hr when very inactive. This heat can add up rapidly in an overcrowded shelter.

Studies have indicated that 85° F is the "line in the sand" for human comfort in a shelter situation. ***Above this temperature, body temperature begins to rise and bodily function begins to deteriorate***.

The air inside the shelter is also highly likely to become saturated with water vapor. In a basement shelter, the very walls may be dripping with moisture. The higher the humidity, the more heated the people in the shelter will feel.

The problems with heat and humidity are the main reason why you don't want candles burning in your shelter, or a little stove to heat foods. Nor do you want people to exercise and generate more BTUs per hour.

If heat and humidity cause problems in your shelter, two major cooling techniques are available to you in times of no electricity.

One: You can transfer the heat and water vapor to the outside air. One way to do this is to have an inverter and car battery set up to run a small appliance, such as a small electric fan. See:

http://www.chemicalbiological.net/The%20Inverter.wmv

http://www.chemicalbiological.net/Get%20The%20Inverter.wmv

If your battery has run dead, or you don't have an inverter setup, you can use a big square of cardboard as a fan to transfer hot, moist air out and pull in fresher, cooler air. Do not fan rapidly. Hold the cardboard above your head then sweep it down to your legs, pause, then sweep it back up while standing before your air intake area, such as an open door or a large opening that leads directly to the outside. By doing this for several minutes, a great deal of air can be transferred and the shelter will be noticeably more comfortable. Rotate this duty with other shelterees.

Two: You can transfer heat to water, via sweat. When the sweat is evaporated, heat is transferred from the body to the air. Old-fashioned hand fans are a good means of moving air in a limited area. Moving air improves the evaporation of sweat from the body, which in turn cools the body. Stock a good hand fan for each person in your home shelter. See:

http://www.chemicalbiological.net/Temperature%20&%20Humidity%20Can%20Take%20You%20Out%20When%20The%20Grid%20Is%20Down!.wmv

Do not use your inverter to run a small electric fan that blows on the people in the shelter. You cannot afford to stock enough new car batteries for this use. The electric fan is strictly for use in transferring moist warm air out of the shelter and pulling in cooler, dryer air at intervals. Place the fan at the air intake area as above.

Another way of keeping the heat down is to move to a cooler area of your shelter. If your shelter is large enough, this may be possible.

Studies have also found that people can withstand short periods of higher heat much better if they have plenty of water for drinking and for sponging or showering. The government allocated one quart of water per day for a 14-day stay in the shelter in their storage plans. That would not leave anything available for heat-dissipation purposes. To withstand heat of around 90° F, the human body requires from 4 to 6 quarts of water per day. Some of this can be used for sponging off.

In your home shelter, hopefully you will be able to have more water on hand in case of serious heat problems.

Keep in mind that high heat and humidity cause discomfort in the shelterees, which in turn may cause disciplinary problems.

Other Methods Of Keeping The Heat Down

You can see from the above that if heat may be a major problem during your shelter time, anything you can do that does not involve adding heat will improve your shelter life. That is why there is such emphasis on non-cook, non-heat foods for the shelter stay.

You will also need lighting for your stay, and candles obviously won't do since they will add plenty of BTUs to an already overheated situation.

Therefore, you will want to have on hand small LED flash-lights for each person and several battery-powered LED lanterns, usually available at Wal-Mart or one of the Dollar Stores. Hopefully, you already make a habit of keeping plenty of batteries on hand of the most popular types that your equipment uses.

Keep in mind that although LEDs create less heat than normal flashlight and lantern batteries, too many flashlights and lanterns in operation can raise shelter heat levels. Everybody wants a lot of light around in order to feel good, but not at the expense of higher temperatures in the shelter.

We have a feeling that AA and C batteries will be worth more than gold once a crisis that takes out the electricity occurs. Store plenty of them, along with AAA and D sizes. One thing you don't need is for your flashlights to dim and go out and you have nothing left to re-energize them.

Residents of the Gulf Coast areas hit by hurricanes that caused power outages that lasted several weeks learned solar lanterns are wonderful items to have. If you have a sunny spot for them to recharge every day, you can then move them to the shelter or to some dark area that you want illuminated during the evening. You won't want to set them outside because you don't want them being contaminated with fallout, nor do you want to go outside into the radiation field unless it's absolutely necessary. But a sunny window area inside your home, say on a window sill, would likely give them enough sunlight so that they can give you several hours of no-heat light during the hours of darkness.

Other Items You May Need

Seeing that temperature in your shelter is so critical, you will need a good room thermometer to monitor heat buildup. When you see the shelter temperature approaching the danger point, you can take action to mitigate the rise as much as possible.

Be sure you have some whisk brooms and dustpans for use in decontaminating yourself and others, should anyone have to go outside during times of fallout. Since fallout is like dust, a whisk broom over your clothes will usually remove most contamination. Use the dustpan and a regular broom to keep the floor swept up of dust and fallout from outside.

Another set of items you will need for a shelter stay are books, playing cards and board games for entertainment. Most family members have certain books in mind they'd like to read but never seem to get around to actually opening. Have each person acquire and keep a small stack of books on hand, their "to be read" pile, so that if something happens, they have their entertainment for their shelter stay on hand.

Inside your shelter proper, you should store some notebooks and pencils and pencil sharpeners. These are for keeping notes on the radiation readings inside and outside your shelter, and for keeping daily records of each person's radiation exposure. If you have young children, crayons and extra paper provide great entertainment.

Supplements To Store

In tense worldwide situations, in addition to your other preparations, take the time to check on your supplements. During a time like this, make sure you and your family can make up for any nutritional deficiencies in your shelter diet by storing the vitamins and minerals and other nutraceuticals your family takes regularly.

Iodine Tablets

One thing you definitely want to store if you believe a nuclear crisis is coming to America is potassium iodide tablets to protect the thyroid gland after a nuclear incident of any type. You can buy Iodoral tablets in your health food store, or another strong potassium iodide tablet. Several online vendors sell iodine tablets especially for use in a nuclear emergency. Get yours now, because once something happens, as in the case of Fukushima, they sell out quickly and cannot be restocked for weeks or months.

By the way, iodized salt does not contain enough iodine to make it protective, nor do kelp tablets. In an emergency situation, you want potassium iodide tablets in the recommended strengths, so get some for your family now.

Generally recommended dosages are:

Adults:	130 mg/day
Children over 18 weighing over 150 lbs:	130 mg/day
Children 3 to 18 weighing less than 150:	65 mg/day
Infants to 3 years old:	16 mg/day

Many tablets are sold in these doses, thus making it easier to know how to break the tablets for each person. For babies or anyone who can't swallow the pills, crush the dose and mix it in applesauce or chocolate milk.

Some iodine tablets being sold proclaim that they protect you from nuclear radiation, but *this is not so*. They do, however, protect your thyroid gland by filling it up with good iodine. The thyroid gland is hungry for iodine and grabs every atom it can from the circulation as blood passes through the thyroid. If your thyroid is already filled with iodine, the thyroid gland lets the radioactive iodine pass it by, to be eliminated from the body through normal processes, or to decay naturally.

Radioactive iodine has about three dozen isotopes, but most have half-lives so short, they are of no concern. The most dangerous radioactive iodine isotope is I-131, which has a half-life of 8 days. The biggest danger here is when it is breathed in during an incident like Chernobyl, or when it falls to the ground and enters into the food chain.

The people in the vicinity of the Chernobyl fallout were in an iodine-hungry state already, and their thyroids literally stockpiled radioactive iodine when it presented, so much so that a Geiger counter held to their throats would register readings like 35 Roentgens. Their thyroids were literally irradiating their bodies and people nearby!

The greatest danger of radioactive iodine is to children. After an incident, unless authorities are on the ball, cows eat the grass coated with isotopes and the radioactive iodine is passed through to the milk they give. Since children usually drink a lot of milk, they may ingest high doses of radioactive iodine this way.

At Chernobyl, the people were not told to avoid drinking the milk from their cows and goats and not to eat the produce in their gardens. Thus, many absorbed a good dose of radio-active iodine, not to mention other radioactive nuclides, before the authorities issued a warning, and they never gave out iodine tablets. This is why thyroid cancer was such a problem after Chernobyl in Russia and the Ukraine, whereas in nearby Poland, there was no epidemic of thyroid cancer because the government immediately issued iodine tablets to the people.

Other Recommended Supplements

To preserve your overall health during a crisis, we recommend a strong vitamin tablet and a strong mineral tablet, taken separately. Vitamin-mineral combination tablets are usually too weak in minerals to do you much good.

During the stress of confinement and radiation, extra B-complex vitamins and Vitamin C are always needed.

Two other supplements highly recommended during times of radiation danger are Carnosine, a powerful free radical scavenger, especially of the hydroxyl radical; and Super Oxide Dismutase, another free radical scavenger, especially of the superoxide radical. Both these supplements help tame the free-radical damage to cells caused by ionizing radiation.

7: *Who Will Get Into Your Shelter?*

Whenever someone plans ahead, as you are doing, and you have a good shelter planned with plenty of food and supplies stockpiled, word somehow gets out. Suddenly, if something happens, you look outside and see a bevy of automobiles pulling up in front of your house, you will know that word has spread to all the people you did not plan to have in your shelter. True to the entitlement mentality so common in America today, these people think you owe it to them to have them in your shelter.

But don't kid yourself that they brought their own supplies. If they didn't plan their own home shelter, we can tell you, they also didn't stockpile any supplies. But you have supplies, and they think they'll just share yours. Or they haul over a grocery sack full of unsuitable items that require heat or other processes that may not be available during this time.

The only way to prevent this scenario is: **Keep Your Mouth Shut**. Tell your children, also. Otherwise, we can tell you that your children's friends will tell their parents, and these parents are all too likely to pack up their family and come to your place, hoping for a spot in your shelter.

You will have no choice but to turn them away. If you don't, unless you are wealthy enough to have a large shelter and a warehouse full of food and water, they will run through your supplies like a buzz saw through balsa wood. Before you know it, the month of food and water you put up for your family will be gone—very likely within a few days.

This happened to a couple during the aftermath of Hurricane Katrina. They happened to have something like six months of food stored, so they told a friend to come stay with them. The friend came, and he brought with him another three couples who were good friends of his. That six months of food lasted less than one month.

Suppose you have the room and enough storage so that you'd like to invite some relatives who don't live with you, or maybe your grown daughter and her husband and children you want to shelter with you.

Or suppose you have some really good friends you'd like to invite to shelter with you in case of a nuclear event. What should you look out for, and whom do you NOT want in your home shelter?

Watch Out For The Macho Types

One kind of person you don't want with you is the person who always knows better than you what to do in every situation. He must be in command at all times, even when he knows nothing about the situation, and the shelter and supplies belong to you.

If you serve peanut butter to go with the survival crackers, he will inform you that soy butter would have been better (never mind that it is feminizing!), and if you tell the people how to properly use the emergency toilet facilities, he will loudly proclaim that doing things another way entirely would be better.

This macho spirit does his best to undermine your authority and to get the other people in the shelter on his side. He grouses about the food and accuses you openly of hiding all the "good stuff" for your own consumption, and he derides all your rules designed to make shelter life easier and more sanitary and pleasant.

Before you know it, you are likely to be tossed out of your own home and your own shelter if he can manage to foment a mutiny.

So if one of these types wants admittance to your shelter, or if someone brings him along and you think you detect this spirit in him, your best bet is to be "filled up" and unable to take in another soul. Because if you let him in, you will have nothing but trouble, guaranteed.

Watch Out For The Vegans and Vegetarians

Why people on any kind of vegetarian diet seem to think everybody ought to be on their diet is one of those questions nobody else can answer, but they can: You should be on their diet because it is right and just and holy. As anyone who has ever encountered vegans and vegetarians can tell you, these are people who have more or less substituted their diet for their religion, and they will bore on about it for hours.

You would think that in an emergency situation, they would eat what is set before them and be glad to get it and keep their mouths shut, but that is not the way it is. They simply must tell everyone why what they are eating is all wrong, and why they should immediately take the vegetarian pledge.

Usually the reasons for their "commitment" to such a diet include such nebulous things as "save the planet" or "stop greenhouse gases" or "save the animals because they are sentient beings." Some will even object to foods that contain eggs or milk, and they can inform you of the reasons in language usually reserved for calling out adulterers and other sinners from a pulpit.

Oddly enough, some people are exactly the opposite of vegans and vegetarians. They are "meatarians" who basically eat nothing but meat. If you serve a meal of vegetable and meat combined, they will pick out the meat and leave the vegetables then grouse because they are still hungry. These people are extremely expensive to feed and are as volatile as the vegan bunch when it comes to their food. If you have both a vegan and a meat-only eater in your shelter, you can look for violent encounters and constant grousing about the food.

Then there are the Paleo diet enthusiasts, or the Adkins diet enthusiasts, or the gluten-free diet , or any other diet that relies on foods you will likely not be able to serve under shelter conditions. On the high-carbohydrate diet of the shelter, they will likely fancy themselves ballooning in weight even though they are not getting many calories—if they eat the food. It's the idea that will get them, and they will make everybody around them miserable.

The same is true of people who are on special diets because of a medical condition. They will often sit proudly, not eating what is offered, in hopes that everyone will notice and that something more "fitting" for their condition will be offered.

The worst is that it is almost hopeless to expect these types to make up their own food storage of foods they can eat. It doesn't seem to occur to them because they expect everyone else to accommodate them.

For obvious reasons, unless you have planned ahead because you know one of the people you want in your shelter is on a special diet, you do not need someone like this in your shelter. They will either be resentful and angry that you can't serve them what they need, or they will be pitiful and hungry. Either way, they will create trouble in your shelter at a time when you least need that kind of trouble.

Now that weight loss surgery has become so popular, you may wind up with a relative or a friend in your shelter who has had this surgery. These people will be much like those on special diets for medical conditions. They need certain foods, usually very high in protein, and most of them refuse to take pills because of digestive problems directly resulting from the surgery.

They may or may not have enough sense to bring their special supplements with them, and if they don't, you can bet they will carry on about the lack of protein in the foods you have put up for this time. Or they will become convinced that during the relatively short period of time they remain in your shelter without their special protein supplements, they will somehow wither away to nothing.

For this reason, we recommend that you think two or three times before inviting a person you know is on a special diet of some type into your shelter. Even if you warn the person to bring his or her own supplies, chances are they just ran out at home and now cannot replace their supplies. Many people refuse to believe anything can happen in America that will interrupt the just-in-time supply of foods and other goods they depend on. But they do believe that if they stock up on supplies, that action will somehow bring on the crisis.

What About People On Prescription Drugs?

People on prescription drugs seem to compose most of the American population today. If they aren't on antidepressants, they're on blood pressure meds, and failing that, cholesterol or diabetic medications. Often, they are on every one of these types of drugs.

Some of these people are perfectly normal people ... so long as they get their medication. If they don't, it's Katy-bar-the-door, because their chemistry may go totally haywire without the medication that keeps it normal.

Other medications, even perfectly ordinary things like over-the-counter allergy medicines, may be the kind of drug that you can't just quit. Horrendous side effects ensue if the person did not prepare with an adequate supply of the medication.

What's worse is that many people know nothing about the medicines they take; they are just blindly following the doctor's orders. When they run out of their medicine and no more is forthcoming, the withdrawal effects can be worse than the condition the medication was supposed to treat.

If you have someone coming to your shelter that you know is on prescription medication, make sure they bring enough medicine to sustain them throughout their stay.

Some medicines are controlled substances, which makes it very hard for a person who relies on these medications to get ahead on them enough for even a short shelter stay. Pain medication is probably the most common of these, and you may have a really hard time if someone in your shelter suffers awful pain because he can't get the needed drug. Aspirin may help to some degree, but a person in really serious pain won't get much relief from anything except the powerful medications their doctors have them on.

Problems will definitely ensue if someone comes to your shelter who is on antipsychotic drugs. It's almost a given that they will not have enough medicine for the entire shelter stay, as many either can't get ahead on their medicines or don't know to get ahead. Often, they think they really don't need the medication anyway, and they don't like the way it makes them feel, so they may even be happy to run out of it.

You haven't seen trouble until you deal with something like this. You may be able to mitigate some of the problem with high doses of B vitamins and other nutraceuticals, but you'll still have trouble getting them to take what you suggest.

Think twice or three times before inviting a person on antipsychotics into your shelter unless you know they have a good supply of their medicine on hand. Dealing with a normal person with weird ideas is hard enough, but dealing with a not-so-normal individual who has run out of the medicine that keeps him more or less normal is darned near impossible.

What About The Fitness Freaks?

We are all for exercise, and we can understand the jittery feelings a person who is dedicated to a daily routine of jogging, weight-lifting, Zumba or some other form of physical fitness regime will feel at not being able to maintain their fitness. Exercise improves the mood, and weightlifters especially enjoy getting their "pump." In a shelter situation, they can almost feel their prime conditioning fading away, and they'll absolutely hate it. Many will succumb to a debilitating depression in this situation.

However, during the relatively short time they may need to spend in a shelter, it will be imperative that they remain as inactive as possible. For one thing, the foods they depend on to maintain their condition will not be available. For another, if they try and maintain some kind of fitness by exercising, they will run up their hunger levels, and they will run up the shelter temperature in the form of heat that they give off, not to mention the moisture in the form of sweat.

If you have a really dedicated exercise freak in your shelter, the person is likely to complain constantly about the boredom and lack of activity, even after you've explained the necessity of remaining quiet and relatively inactive.

Unless you are very certain this person will understand the reasons for inactivity during the shelter stay, think twice about inviting to your shelter someone who absolutely cannot do without exercise.

What About Your Grown Children?

Most parents, in spite of evidence to the contrary, persist in believing they know their grown children. Often, once away from home, your son or daughter subscribes to entirely new ideas and beliefs that may be diametrically opposed to what you taught in your home.

Also, your children, male or female, may come under the influence of the person they marry, and that person may hold ideas that you can't understand or agree with. Thus, if they come to stay in your shelter for the duration of a fallout situation, you may find yourself with two people you don't know in very close proximity.

Unless you are certain they will follow your rules while in your home shelter, it may be far better for you to let them go elsewhere rather than invite them to your shelter.

What About Pets?

Your own pets are members of your family, and if you want to have them in your home shelter with you, that is certainly your prerogative. After all, you have stored up enough of their favorite foods to last them through the shelter stay, so they are accounted for and welcomed.

But what about any person you invite to your shelter who has a pet?

You will have to make it very clear that pets are not allowed, no matter how well-behaved. If a person has a pet they're very attached to, they will either go elsewhere, or they will find other accommodations for the pet.

Again, this is a situation where you may not even know about the pet until the person shows up, pet in tow. You may find they expect you to let the pet stay in the shelter with them, and that you feed the pet its favorite food, which they forgot to bring along. Also, pets pose yet another sanitary problem, as you will not want them going outside and coming back in coated with fallout.

And don't forget, many people are allergic to animal dander, which may cause a serious problem for some of your shelterees.

If you know a friend or relative has a lovable pet they are extremely attached to, think twice before you invite them into your shelter. This is more trouble of a kind you don't need at this particular time.

Remember, in a nuclear situation where fallout requires you to shelter, money will be worthless, even gold and silver as the Bible stated. Nothing will be moving, including delivery trucks, and nobody will be selling.

Some individuals will offer to pay you for a place in your shelter, but even if you are well-stocked, you still cannot accommodate people with specials needs and special diets, not to mention prescription drugs they run short on.

These people may even promise that their special requirements will make no difference in their shelter behavior, but *you now know better, because you are reading this book.*

Once their brain chemistry becomes deranged because of a lack of their medicines or whatever else they depend upon, they may not be able to help themselves.

8: *Your Shelter Rules*

What should you do if you have an excellent shelter area and you would like to invite certain people to shelter with you, especially some of the relatives you respect?

We recommend that **first**, you examine your reasoning thoroughly and make sure none of the people you want to invite are types you know will cause trouble. You may have a close relative who is a macho spirit, but *do you really want him in your shelter* under trying circumstances?

Second, you might want to discretely check on their medical status in terms of prescription drugs or special dietary needs. You can avoid a lot of trouble by making certain the person has no special needs or psychological problems that would make for trouble in the shelter.

Third, if you are determined to invite some close friends or relatives to shelter with you, we suggest that you first give them a list of *Shelter Rules* that spell things out in no uncertain terms. Pay close attention to how they respond to the rules, especially their expressions as they read the list. Otherwise, you are sure to end up with the kind of trouble you don't want.

Many people simply do not understand what a fallout shelter is all about. Sure, it protects them from fallout, but the idea that they might not have everything they want to eat, and that there might be no entertainment in the form of television and their favorite music or video games just has not occurred to them. The fact that there may be no cell phone service, or that they won't be using their cell phones in your shelter except one or twice daily if the phone service is still up, may not be fathomable to some people.

That is why you would be wise to spell these facts out. Assume the people know nothing about radiation or the circumstances likely to be in existence after a nuclear attack and lay out the probable conditions they may find. Then be sure and inform them that if they have special dietary needs or require certain prescription drugs, to stock up on at least a month's supply of the things they require to maintain their health.

One way to search out people's thoughts about this is to bring up the subject and ask what their plans are if Russia, China, or some rogue state decide to land a few nukes on American cities.

You will learn, hopefully not to your surprise, that many people do not even know that public fallout shelters are no longer in existence. *The idea that they will have to do for themselves may come as a rude shock.*

Of course, it is possible that FEMA would remain viable and might open a public shelter, but if you remember anything about the Katrina and Rita shelters, you do not want to be in one, and it's likely your friends and relatives don't either.

The fact, however, is that many people now are like sheep. If the government says to go to a FEMA shelter, these people will march off in neat lines to a FEMA shelter. And it likely still won't occur to them that they'd better bring along a supply of their prescription medication and any special foods they must have.

After you explain to them what is needed for a fallout shelter, suggest that they read this book, or Cresson Kearney's book, or any of the Civil Defense publications on the subject from the 1950s and 1960s.

For these publications, go to:

http://www.chemicalbiological.net/Downloads.HT ML

Before you invite them into your shelter, make sure they know a little about the subject. Then your rules won't come as an unwelcome surprise.

If you find that one or two of the individuals you would like to invite to your shelter have no intentions of obtaining a supply of the medications or foods they need, you would be most unwise to invite them to your shelter. This is a person who does not want to have to do anything for himself, but if he winds up in your shelter, he will make a nuisance of himself complaining that you didn't take his needs into consideration.

Consider yourself warned. This is why you really do not want others in your shelter except your immediate family.

The same is true of any person you love and would like to have in your shelter, but who has a pet, special needs, is a vegan, or a Paleo diet enthusiast, or has any other problem that requires special treatment. In a shelter situation, many of these needs, even though they may be required for that person, simply cannot be met unless the person himself plans ahead for his or her own needs.

That being said, when the world heats up and you believe it may soon progress to all-out nuclear war, that is the time to invite those you have chosen that you want to shelter with you.

Once they get to your shelter, you will give each person a copy of your Shelter Rules. But before the necessity arises and they accept your invitation and come, here are what we

81

call the Preliminary Rules that people must understand *before* they come to your shelter.

The First Rule, and one that you had better make sure is explicitly laid out and understood is: ***Do Not Bring Anyone With You***. There Is No Room At The Inn!

The Second Rule is that if a person has special needs in terms of diet or medications, he has the responsibility to provide these for himself, as you would have no way of doing so.

The Third Rule is that if they have pets, they must make other arrangements for their pets. If they feel they cannot be separated from their pet, then they had best set up their own home shelter.

The Fourth Rule is that radios and I-pods are not allowed at a time like this. No one wants to listen to anyone else's noise. Cell phones will likely not be working after an attack, and if they are, they may not be rechargeable if the electricity is out.

The Fifth Rule is that you must bring your own reading matter or crossword puzzles or whatever *quiet* entertainment you enjoy.

The Sixth Rule is that everyone should bring a change of clothes and socks, along with a whisk broom and small dust pan. This is in case they are caught in fallout and need to decontaminate themselves and change into uncontaminated clothing before coming into the shelter area. If they come to your shelter after fallout is already falling, they should protect themselves by wearing a big hat, gloves, boots, and a raincoat. Then they can shed these outer garments in one of the outer rooms before coming into the shelter.

The Seventh Rule is that there will be no talk about what food is served, or what foods they wish were being served. Talking about food brings images and desires into peoples' minds, and the next thing you know, there will be grumbling and complaints about what is served. This will generate a severe discipline problem!

The Eighth Rule is that there will be no exercise during the shelter period. Everyone will rest quietly and read or pursue other quiet activities, as calories and water may be very restricted and nobody needs heat or humidity buildup to become a problem in the shelter.

The Ninth Rule is that there will likely be no capacity for excessive personal care. In other words, people will not be able to do more than sponge off, assuming the water supplies hold up. The problem will usually be caused by people who have dreadlocks, braids or some other hairstyle loaded with lots of hair pomade or hair oil who are out in active fallout and get it in their hair. In a situation like this, there will be no way for them to shampoo their hair and get rid of fallout that could easily be brushed from normal hair. If there is someone who favors a hairstyle like this that you'd like to invite to your shelter, bear this problem in mind. Also, all shelterees should avoid the use of perfumes.

The Tenth Rule is that anyone with a baby must bring all the required baby supplies, such as diapers, wipes, formula, baby food, etc.

The Eleventh Rule is that everyone coming to your shelter must bring a box of white plastic garbage bags (13 gallon) and a box of Wet Wipes, preferably two boxes of each. The bags are useful for many shelter operations, and it would be hard for you to store as many as you would likely need in a case where other people come to your shelter. The same is true of Wet Wipes. In a situation with no running water, people will use plenty of them.

The Twelfth Rule is that there can be no smoking. People can't be sneaking outside for a smoke, nor can they smoke inside the shelter or your dwelling. No matter where they go inside to smoke, other shelterees will be forced to inhale some of their smoke. If they are always going outside, they will be out in the fallout unnecessarily and bringing it back in with them. If you don't smoke, it is best not to have a smoker in your shelter.

Once each person comes into your shelter, give them a copy of your "Shelter Rules." The purpose of these rules is to make the shelter stay easier on everyone and to make sure that there are no misunderstandings.

An example of such Shelter Rules follows:

Shelter Rules

1. Bring no one else with you to the shelter.
2. Bring all special foods and medications you require.
3. No pets are allowed.
4. Radios, I-Pods and other noise-makers are not allowed.
5. Bring your own reading matter or other quiet entertainment items.
6. Bring a whisk broom and a change of clothes in case your clothing should get contaminated.
7. No discussion of what food is served or what you wish was served. Eat what is set before you without complaint!
8. No exercise allowed during the shelter period.
9. There are no facilities for excessive personal care.
10. If you have a baby, you must bring all required baby supplies, such as diapers, formula, baby food, etc.
11. Bring with you 2 boxes of 13-gallon white garbage bags and 2 boxes of Wet Wipes.
12. No smoking, drinking, or recreational drug use.

To Avoid All These Problems Have No One In Your Personal Shelter Except Your Immediate Family, Who Have Been Trained By You.

An additional consideration, depending upon your level of preparedness and your ability to teach others, is a rule requiring each person coming to your shelter to bring along their personal protective (gas) mask, outfitted with oxygen. After a nuclear attack, out-of-control fires are likely to rage all around, filling the air with poisonous gases, ash and soot, not to mention radioactive particulates. Oxygen may be in short supply in the surrounding air.

It goes without saying that most people will not have a gas mask, nor will they see the need of one ... until they look out on a landscape of raging infernos, and even then they may not see the need of a gas mask or oxygen. By then, it will be too late to obtain either.

9: *Getting The Shelter Ready*

When you notice a worsening crisis in the news, it is time to begin checking your supplies, testing your radiological meters and zeroing your dosimeters.

Once you hear that there has been even one nuclear explosion in the United States, assume there will be more and immediately construct your expedient shelter and/or ready your shelter for occupancy.

After even one nuke in the United States, it is highly likely that no trucks will roll, and the grocery stores will likely be emptied out within a few hours. This is the main reason why you should keep your battery and food supplies built up by immediately replacing anything you use.

The first thing to do is shore up your shelter physically, if necessary. Check for any weak areas or holes that radiation could come through and pack more totes with books or other material to shield those areas. If you have a window to the outside that you will depend upon for ventilation, make sure it is as shielded from fallout as you can make it.

Charge the automobile batteries that will run your electric fan with the inverter if they need charging. Zero your dosimeters and do operational checks on your radiological meters.

If more bombs are exploded, it is time to actively prepare for fallout. Immediately fill your bathtub with water and fill every available container in your kitchen with water. If the situation gets worse, there may be no electricity and no running water very soon.

Check your kitchen cabinets for items you regularly use in cooking to see what can be used to spice up your shelter food supply. If you are able to cook in your kitchen, you can often plan a more rounded-out diet than what you would have if circumstances are such that you could not cook.

Check your writing supplies to make sure you have plenty of paper and pencils on hand for recording daily radiation doses and to keep radiation logs of readings inside your shelter and at selected places outside your shelter.

Now would also be the time to choose places to hang dosimeters in order to keep up with the dose in certain areas of your dwelling.

Make sure all your windows, doors, and any other openings to the outside are closed if fallout is expected.

Another way to monitor for fallout is to place white paper plates in various locations around the outside of your dwelling so that you can watch them from the windows. Once you start seeing what looks like dust or sand on those paper plates, you will know fallout is arriving.

By this time, you should have your meter ready. Start taking readings at various areas around your house, and even outside if your paper plates remain clean. Once fallout starts coming down, you do not want to be outside. These early readings will also tell you what normal background radiation in and around your dwelling is.

Or you can use the following technique:

http://www.chemicalbiological.net/Dosimeter%20Reading%20Converted%20Into%20Dose%20Rate.wmv

If you have a CD V-717 meter with the 25-foot distance capability, you can set the detector section up outside so that you can monitor radiation levels on the dial section from inside your home. For a short video on how this works, see:

http://www.chemicalbiological.net/CD%20V-717%20Field%20Trip.wmv

Ideally, before fallout begins arriving the people you invited to shelter with you should pack up any supplies they will need and come at once to your house. Gather your family members and have them prepare to enter the shelter area.

Once fallout begins coming down in earnest, and you start getting radiation readings on your meter, everyone should enter the shelter area and remain there.

As the shelter manager, you should take readings every few hours while fallout is coming down so that you can monitor the rise in radiation levels. Keep a careful record of the radiation levels and where in your dwelling the readings were taken. This information will be valuable later in helping you to choose the safest areas of your dwelling and the safest areas of your shelter.

Everyone should have on a CD V-742 dosimeter, provided you have enough dosimeters on hand. Be aware that some of your shelterees may try to keep the dosimeter. Do not allow this! You may need all your dosimeters again in the very near future.

10: *How To Run Your Shelter*

Once fallout begins arriving, one of your primary duties will be to monitor radiation levels, both in your shelter and in the area surrounding it. These levels are important because they will tell you if your shelter is protecting the occupants well enough, when you will be able to leave the shelter to do certain errands outside, and most of all, when the levels of radiation will have decayed enough so that the occupants can safely leave the shelter.

In order to perform radiation checks, you must have a radiological meter. If not, see Book I of our Radiation Series, *How To Choose A Civil Defense Radiological Meter: Geiger Counters & Dosimeters.*

If all you have are dosimeters (CD V-742) and a dosimeter charger (CD V-750), you will be able to keep a fairly good record of radiation dose rates and doses, but you will not be able to check people or foods for contamination. Dosimeters, especially the Civil Defense dosimeters, are generally designed to read the higher levels you would receive in a nuclear incident or in a fallout zone. They would be of little use in detecting the much smaller amounts of radioactive contaminants in foods or on people.

However, if you have enough dosimeters, issue each person who enters your shelter a dosimeter and have them sign for it. Record the serial number of the dosimeter on a sheet of paper that has their name on it. Be sure and check the dosimeter back in when the shelter period is over. You do not want to lose any of your dosimeters.

Tell each person to clip the dosimeter to their pocket or at their waistline. Every night, you or the person will fill out a sheet for that person, listing the dosimeter reading for that day. You can either re-zero all the dosimeters, or subtract the reading from the next day's reading to keep a record of each person's daily dose of radiation. These records may be very important in the person's healthcare in the future.

When Fallout Begins Arriving

When you receive word that fallout is arriving in your area, assuming communications are still up, or if your meters detect radiation, that is the time to begin taking readings—at the first sign of radiation arrival. Before this happens, you should have taken enough readings to have an idea of the normal background radiation in and around your dwelling.

How To Use Your High-Range Meter

With the CD V-715 (or 717 or 720) warmed up for two to three minutes and zeroed, go to the area where you want to take a reading. Holding the meter at waist-height, about 36 inches off the ground, observe the scale. If the range selector is set at X0.1 and the needle goes off the scale, switch to the next range, the X1 range. Do this until you find the range that allows you to obtain a good reading. Generally, when fallout first begins coming down, the X1 range is the most suitable range, but this depends upon how heavy or how light the fallout in your area is.

If you have trouble obtaining a good reading on the X0.1 scale because the reading is so small, get out your CD V-700 and take a reading. Remember, even though the readings may start out small, if fallout continues, the radiation levels will climb steadily and you may soon have to go back to your CD V-715.

Once you have obtained a reading for an area in your dwelling, record it, along with the time and date, in your Shelter Records.

Using your CD V-715, check your dwelling and your shelter area, taking readings at the spots you have decided to use for your daily radiation readings, and record them either on a sheet of paper or on a drawing of your dwelling.

Checking For Safer Areas In Your Shelter

Take these readings in the same places every day, several times a day while the radiation levels are rising. You must find out what areas of your dwelling and shelter are more protected than others.

If your shelter is large enough, say a room-sized area in your basement, it is important to take readings in various areas of the shelter. You may learn that certain areas of your shelter are more protected than others because you obtain consistently lower readings in those areas than in other areas of the shelter. These protected areas are the locations you want to tell your shelter inhabitants to stay in as much as possible.

People should, if possible, occupy only areas with a dose rate of less than 2 Roentgens per hour. If this is not possible, then occupy the areas with the lowest readings.

After taking your readings, note the protected areas in your original drawing and record the readings you took at each spot in the shelter on that original drawing. This will help you determine the safest areas of your shelter, and may give you hints as to why those areas are safer. You may be able to fortify those less safe areas in some way and increase your safety.

Record all these readings carefully, including the day, the time, and the location. After the emergency is over, these readings will help you determine what changes you may need to make in your shelter, and what doses of radiation each person in your family may have received.

Rotational Method of Radiation Dose Control

If your shelter is large enough, you may discover that there are certain areas of the shelter that have better protection than others. This is not a problem so long as the readings differ by less than 10 Roentgens.

If there is a difference of 10 Roentgens or more between one or two areas of your shelter, and the shelter is occupied in full, then you will need to rotate the people in order that no set of persons receives a higher radiation dose than the others.

Depending upon how many people are in the shelter, and how much room you have, you can label the areas with letters such as (A), (B) and (C) and divide your people into three groups also. Allow each group of people to stay in each area for 1 to 2 hours before moving to another area.

If you have children or a pregnant woman in your shelter, however, keep them in the safest areas of the shelter at all times. They are far more susceptible to radiation damage than other adults.

If there is room, it is possible that you can crowd people into the safest areas for reasonable periods of time. The important thing is to keep each person's radiation dose as low as possible.

Checking People For Contamination

Some of the inhabitants of your home shelter may not arrive until after fallout is already coming down. If you have warned them of this possibility beforehand, they should know to arrive wearing a raincoat or some other outer garment, boots, and a large hat, and if they have long hair, tuck the hair up beneath a turban or a shower cap. You do not want people coming inside your shelter area with fallout on them that will add to the amount of radiation you and the other occupants of the shelter will receive.

Once the people enter, direct them to an outer room of your dwelling where they can shed the outer protective garments. After they have laid aside their outer clothing and brushed off with the whisk broom they should have brought in a pocket, check them with your CD V-700 low-range meter.

Using Your Low-Range Meter

Your CD V-700 low-range meter is designed specifically to detect low levels of radiation. These lower levels are usually found on people who have come into contact with fallout, or on foods and water which have been out in fallout.

In order to check a person who has just come into your shelter and has shed his outer garments and brushed off, turn on your meter and allow it to warm up for 30 seconds. Open the shield on the "hot dog" probe and hold the probe close to the radioactive check source found on the side of the meter. You should read between 1.5 and 2.5 milliRoentgen per hour.

Generally, the CD V-700 is carried via the shoulder strap over one shoulder, while you hold the probe in your dominant hand. When checking people for contamination, you should wear the headphones rather than use one of the after-market speakers that broadcast the clicking sounds as the meter registers radioactivity. People can get most upset if the sound suddenly accelerates while the probe is being run over their bodies. This can create a panic situation.

When checking people for contamination, you should put a plastic bag or a thin sheet of plastic around the probe to protect the probe from becoming contaminated itself.

Have the person you are checking stand with legs slightly spread and hands held out about a foot from their sides. Run the probe over all areas of the person's body, covering about 1 or 2 inches per second, and taking care to especially check the most likely areas for contamination, such as feet, hands, and shoulders. Do not touch the person with the probe. Keep the probe about ½ to 1 inch from the person's body.

Listen for an acceleration of the clicking noise, and keep an eye on the meter scale as you do this. Contamination is generally defined as a reading of about 10 milliRoentgen above normal background radiation. Some experts call *anything* above background radiation contamination.

For a visual demonstration, click on this link:

http://www.chemicalbiological.net/checking%20for %20contamination%20using%20a%20GM%20meter .wmv

If you discover someone's body has picked up some fallout contamination, have them brush those areas again with their whisk broom and use waterless wipes to cleanse the skin or hands. Then check the person again until they test free of contamination.

Checking Food For Contamination

Foods and liquids that are already inside your dwelling should be safe from fallout contamination. Canned goods in your cabinets, foods in your refrigerator, water in your storage barrels ... all these are considered safe. The only foods and liquids you should test for contamination are foods brought in from outside the shelter after fallout began.

Even then, food inside wrappers or cans is usually not contaminated. All you need to do is wipe down the outside of the container or wrapper then open the wrapper or can and extract the food without touching the outside of the container.

If possible, you should avoid eating produce from your garden or drinking milk from your own cows if they are grazing on grass that has been contaminated with fallout. After waiting a few weeks for the fallout to decay to safe levels, these foods are often safe again.

If you need to check some food items for contamination, perform the check the same way you would on a person. You are testing for outside contamination only. To test a food such as fish, where the animal has taken in the radiation, generally requires laboratory equipment you will not have available in your home shelter.

Open the window on the probe of your CD V-700 then cover it with plastic. Slowly run it over the surface of the food item you are testing, about ½ to 1 inch above the surface. Again, you are looking for an increase above background radiation. Some authors specify a reading of 10% above background radiation. This is why it is important to know the usual background radiation readings in your home.

The Protective Factor Of Your Shelter

The Protective Factor of a particular shelter is a measure of how much protection the shelter provides from radiation. In general, the Protection Factor of automobiles, houses and strip malls are less than 4, which is considered *Poor*.

Two-, three- and four-story homes and apartment buildings are considered *Inadequate*, with Protection Factors of between 4 and 9.

Residential basements, the best locations in 3-story brick apartment buildings, the outer areas of mid- to high-rise buildings with brick or concrete walls are usually *Adequate*, with Protection Factors between 10 and 39.

Large basements or underground areas and the inner areas of mid- or high-rise buildings with brick or concrete walls are considered *Good*, with Protective Factors of 40 or above.

You will also want to know the Protective Factor of your shelter, as it will come in useful later. In order to find this, you will need to take a reading outside, about 25 feet from your dwelling. If fallout is still coming down, wear a big hat, boots, and a raincoat to make sure no fallout gets on your clothing or person, and wear a respirator to keep from breathing in any radioactive particulates. Cover your CD V-715 with plastic. Take the reading as quickly as possible, then get back inside your dwelling.

Shed your outer clothing and brush off any areas that might have collected any of the radioactive dust, remove the plastic from your meter, then go to your shelter. Take a reading inside your shelter.

The Protective Factor of your shelter is calculated by dividing the outside reading by the inside reading of your shelter. The answer is your Protective Factor.

Suppose you obtain a reading of 20 Roentgens outside your house, and inside your shelter, you obtain a reading of 2 Roentgens. Then:

$$20/2 = 10$$

The Protective Factor of your shelter is 10. This means that if you obtain a reading of 1 Roentgen inside your shelter, then you can multiply that reading by 10 and it will give you the probable reading outdoors: 1 x 10 = 10 Roentgens outdoors.

Knowing your shelter's Protective Factor means you no longer have to go outside into the radiation to get an idea of what the radiation readings outside your dwelling are. Periodically, you may want to go outside and take a reading, just as a check, but in general, your best bet is to stay inside your shelter and use the Protective Factor to get an idea of the readings outside.

Maintaining Daily Dose Records For Each Person

Your shelter records should contain a page for each person, on which, each day, you enter the date and the dose of radiation the person received on that date. This is where your dosimeters come in. You can use a notebook for this purpose, or you can print out multiple copies of the chart found at this link:

http://www.chemicalbiological.net/Radiological%20Meter%20Page!.html#Radiation Dose Record

Be aware that keeping daily records of each person's exposure can prevent panic. People can literally scare themselves into believing they are dying of radiation sickness, so much so that they even manifest symptoms. If you can show them that their dose readings are too low for symptoms, the situation will usually calm down.

Radiation Dose Record
For

Name of Shelteree

Time And Date	Added Exposure	Total Exposure To Date	Comments

A Radiation Dose Record Chart

To see an example of a chart in use, go here:

http://www.chemicalbiological.net/Radiation%20D ose%20Record_1%20and%20_2.html

If each person is wearing a dosimeter, an ideal situation, you can read the dosimeters at the same time each day. Enter the reading on the person's sheet. Then you can either re-zero the dosimeter, or you can subtract today's reading from yesterday's reading to arrive at the dose received per day.

Using Your Dosimeter To Calculate Dose Rate

When you take a reading with a meter, you are reading the *dose rate*, which is how much radiation is coming in per hour. We usually read the dose rate as Roentgens/hour, and in order to find out how much radiation (dose) we have received, we multiply the dose rate by the time we were in the radiation field.

For instance, if you were outside for 30 minutes, and your CD V-715 tells you the radiation dose rate is 60 Roentgens/hour:

$$60 \text{ R/hr} \times 30/60 = 30 \text{ R}$$

In ½ hour in a radiation field of 60 R/hr you would have received a *dose* of 30 Roentgens.

A dosimeter, however, only reads *dose*, or how much radiation you have absorbed over a period of time. So if you know the amount of time it took you to receive the dose recorded on your dosimeter, you can arrive at a good estimate of what the *dose rate* was.

Suppose, as in the example above, you were outside for 30 minutes (1/2 hour), and your dosimeter, which read 0 when you went outside, now reads 30 R. Then if you divide the reading by the time you were outside (in hours), you can arrive at the *dose rate*, or how much radiation was arriving per hour.

$$30 \text{ R} / ½ \text{ hr} = 30 \times 2/1 = 60 \text{ Roentgens/hr}$$

This calculation will not be as accurate as a reading taken with a radiological meter, but it will give you some idea of how strong the radiation field is.

Finding Protection Factor of Shelter

To find a Protection Factor in your Home Shelter: 30 minutes after detonation, radiation levels are shown to be around 700 R/hr outside. If you found the lowest level in a certain area of Your Home Shelter is about 70 R/hr, Divide Outside Reading by The Inside Reading To obtain the Protection Factor (PF), which will be equal to 10.

Take The Outside Reading And Divide It By The Indoor Reading:
700R/hr Divided By 70R/hr = 10
Which is The Protection Factor of Room

11: *Managing Your Shelter*

Managing a fallout shelter, even if the shelterees are your own family members, is not for the faint-hearted, and managing a shelter composed of friends and more distant relatives can be even tougher. Part of the reason is because people now do not understand the necessity of constant self-discipline in a situation like this. Nor can they change their behavior, because they have never tried.

People of baby-boomer ages and younger have never lived through really tough times, have never lived through war conditions where food rationing was imposed and have never experienced what it was like not to be able to run out and buy what they needed when they needed it. For this reason, even though their heads tell them that your strictness about certain things is necessary, their "free-spirit" mentality may cause them to object.

Your best means of handling objections is to patiently explain the reason behind them. If necessary, you can remind them that *this is your shelter and your rules are the ones that apply here*. If they can't follow your rules, they may need to brave the radiation and get back to their own homes to quickly set up their own home shelters.

Communication

Many people will try and rely on government information during a nuclear crisis in the mistaken belief that the government knows what it's doing and will tell the citizens the truth about the situation.

In reality, those days are long over. In spite of studies showing that people in a crisis situation perform better when told the truth, governments can usually be counted on to lie in order to (1) look like they know what they're doing and (2) keep the people from panicking or otherwise taking action on their own.

In your own home shelter, if you have the necessary equipment in the form of a radiological meter or two and some dosimeters, you can discover the truth for yourself and tell it to your shelterees.

You may not know the directions fallout is traveling and what areas can look for fallout to arrive, but you will know when it does arrive in your area, because your meter will detect it. And if you know the direction the wind typically blows in your area, you can likely make a good guess as to where any fallout is going if you know which areas were bombed.

We advise that you make regular announcements to the people in your shelter, telling them what the radiation readings of the day are, and how they appear to be either increasing or decreasing. Assume that your people are intelligent beings who appreciate the truth, because it is unlikely you would invite anyone to stay with you who prefers not to know the truth.

Dispensing Water Rations

One of your major jobs as Shelter Manager is to make sure the people in your shelter receive their proper allotment of food and water every day. If you have more than three or four people in your shelter, you may wish to designate someone to assume this duty, or to help you with it.

Depending upon your storage capacity and the amount of time you estimate your group may have to remain in the shelter, you can decide on how much water each person gets during a day. Each person needs at least one quart per day. Two quarts would be better.

Public shelters planned to deal with the dispensing of water by allotting each person in the shelter a plastic cup. That cup would be filled at several specific intervals throughout the day.

In your own shelter, depending upon how many people you have in the shelter, a better way might be to give each person a quart container and a cup. The quart container is filled once (or twice, depending) each day, and the person can pour out his own amounts throughout the day, with children under the supervision of their parents.

Dispensing Food Rations

Food is serious business in a shelter situation. Depending upon the circumstances and whether or not you will be able to cook or heat foods, seeing that your shelterees receive their daily allotment of calories can be a very big deal.

Remember that the people in your shelter, including you, are all worried about what's going on in the world, what's next in the nuclear war, and what's happening to their friends and relatives who are scattered all over the country. In addition, their current situation involves a lack of privacy, a lack of customary activities and a lack of fresh air and freedom. Everyone in the shelter must exert great self-control and exert themselves to maintain strict sanitary practices and careful use of available food and water.

Considering that boredom and inactivity can be expected to cause problems in your shelter, one way of mitigating this is to give each person certain responsibilities. But be sure you choose the most suitable person for each responsibility, and be sure each person can be trusted to follow your rules.

The big thing when food is strictly rationed is to make sure that there is no perceived favoritism when handing out servings. If, for example, you choose a woman who has one child with her in your shelter to hand out meal servings, you can bet that everyone else will be watching what she serves her own child. Heaven forbid that people observe her child receiving so much as one extra survival biscuit more than everyone else. You will likely have a shelter mutiny on your hands.

If you are able to cook a little in your own kitchen, you can often eke out the relatively meager diet of survival biscuits with vegetables, canned meats and casseroles. But if cooking proves impossible, you can still eke out the survival biscuit ration with toppings such as a tablespoon of peanut butter, or a tablespoon of strawberry jam.

When the situation is so bad that you can't cook or heat, you can bet you won't be washing dishes either, because that will take water you don't have. If you don't have a big supply of cheap paper plates, then use napkins or paper towels to serve the biscuits, and place the tablespoon of topping on one of the biscuits.

If you are able to heat water, you can make hot coffee for your shelterees. Many people depend upon their morning cup of coffee to get them going, and most coffee-lovers agree that a cold cup of instant coffee doesn't do them nearly as much good as a hot cup of percolated coffee. But that cold cup of instant is still considered far better than no coffee at all.

During the two weeks of shelter time, unless you have canned evaporated milk on hand, or somebody brings some, there will be no milk for coffee or for children.

Studies on the subject of public shelter management agree that adhering to a regular schedule of meals is best for all concerned. Everyone knows what to expect, and there are no surprises.

Other Aspects of Shelter Management

One of the best ways to obtain the cooperation of everyone in the shelter is to explain the rules and the necessities for them upon their arrival at the shelter.

Explain up front the reasons for food and water restrictions, and the need for everyone to remain quiet and relatively inactive. If it looks as though heat will become a problem, explain that also, along with what you plan to do about it.

If you explain everything that is going on in so far as you can tell, including the fallout readings and what they mean to the people, you will receive much better cooperation. People can often be reasonable if they know the reasons for doing things a certain way.

Food and water supplies, along with medical supplies and your radiological instruments, must be kept in a separate location, preferably under lock and key, and all supplies must be accounted for at all times. One thing you do not need is for some of the people to try and steal food because they think they are "starving." Another is for children to play with your meters. They are not toys.

Wastes and garbage must be kept in closed containers in another room of the dwelling. Before placing it in the garbage container, put garbage into a Ziploc bag or roll it up in old newspaper or butcher paper. Sanitation is going to be a big problem and will wreck a shelter if it is not kept clean.

Strict Sanitation

Nowhere are sanitary practices so important as in a shelter where people may be crowded together. Whenever people are in close proximity like this, it sets up what is called a "refugee situation," where the crowded conditions and lack of sanitation can bring on an *explosion of diseases such as cholera and typhoid fever.*

Strict sanitation basically means human-waste control and insect control. It also means avoiding cross-contamination, where you touch one thing that contains bacteria and transfer that bacteria to something else you touch.

For this reason, disinfectants should be kept in the toilet area and each person must pour a little into the toilet after each use. They should also cleanse their hands after each use of the toilet. In a situation like this, their continued good health will be up to them.

One way of providing a means of cleaning the hands is to provide plenty of wet wipes. You cannot have too many wet wipes in a shelter situation. Clean the toilet with Clorox wipes after each use.

In the kitchen, you must take special care that those who distribute the food and water rations do so only after cleansing their hands, especially under the fingernails. They must also take care not to cross-contaminate, that is, after cleaning their hands, they touch something that isn't clean before handling the food.

Remember! In a situation where people are crowded together for a period of time, sanitation becomes doubly important, especially when there is no running water.

Insect Control

Flies have long been known to be an important vector of disease, in particular typhoid fever. In a shelter situation where there is no running water and no electricity, you must take extra precautions to avoid allowing insects into your shelter.

You should always keep roach spray and flying insect spray on hand in your home, so that when the time comes to activate your home shelter, you will have it on hand. Do not allow flies, mosquitoes or roaches into your home or shelter during this time.

Many homes do not have screens these days since people depend so much upon air conditioning, but no decorator seems to give a thought to what is needed when the electricity is no longer up and the windows need to be opened for ventilation. With this in mind, if your home does not have screens, you might wish to get some made to fit your windows. Failing this, have on hand mosquito netting that you can tack up over windows you decide to open for ventilation.

Flies are attracted to garbage and wastes. This is why it is so important to keep wastes and garbage in tightly covered containers away from the food area. If possible, you may want to move filled garbage and waste containers outdoors to a location some distance from your dwelling.

Every day, the shelter area should be swept, and all trash picked up regularly and placed in garbage cans, after wrapping in newspaper if needed. Cleanliness must be emphasized to all the shelterees, and each must help in keeping the area clean.

Radiation Levels and Outside Activities

No matter how much you store and how prepared you are for your shelter stay, something always crops up that needs doing outside the shelter. The question is should you do it?

Much depends upon what the radiation levels are outside the shelter.

If radiation levels outside are 100 Roentgens or above, do not perform any outside activities. Wait until the levels have declined before venturing out.

If radiation levels are between 10 and 100 Roentgens, only go outside for a very few minutes, and only for a reason that cannot be postponed.

When radiation levels have declined to between 2 and 10 Roentgens, you can be outside for periods of less than an hour and only for essential tasks.

Radiation levels of 0.5 to 2 Roentgens will allow you to be outside for several hours a day for things like firefighting, burying wastes, or obtaining more supplies.

Once radiation levels have declined to less than 0.5 Roentgen, no more precautions are necessary, and you can be outside as much as you need, although you should still sleep in the shelter.

When radiation levels have decayed to background radiation levels, everyone can leave the shelter for good.

12: *Making Your Own Survival Crackers*

Survival crackers, as Civil Defense called them, actually have a long and interesting history when it comes to war. Only the stuff back then was called *hard tack*, and when they said hard, they meant hard.

Crackers are actually a "savoury biscuit of British origin," according to *LaRousse Gastronomique*, and the "manufacturing method gives it a flaky, crumbly texture."

A **biscuit**, on the other hand, is "a sweet or savoury dry flat cake with a high calorie content." The biscuit was so named because it was cooked twice, and this second cooking renders the biscuit "dryer and harder and improves its keeping qualities."

"Keeping qualities" make the biscuit very useful, and *LaRousse Gastronomique* adds, "This very hard, barely risen biscuit was for centuries the staple food of soldiers and sailors. Roman legions were familiar with it, and Pliny claimed that 'Parthian bread' would keep for centuries. Under Louis XIV, soldiers' biscuits or army biscuits were known as 'stone bread.'"

In later years, this biscuit sailed with the British Royal Navy, served as a thickener and key ingredient in New England chowders, and fed soldiers on both sides during the American Civil War (1861 – 1865).

For use on a ship, the biscuit was cooked four times rather than two, then allowed to sit for six months so that it could dry out even more before the ship sailed. As you might expect, the keeping qualities of this particular biscuit were phenomenal.

The *ship's biscuit*, as it was called, became known as *cabin bread*, *pilot bread*, *sea biscuit*, *sea bread*, or *hard tack*. According to Wikipedia, *tack* is a British slang word for food. Moreover, other descriptive terms such as *molar breakers*, *sheet iron*, and *tooth dullers*, tell us why we would be wise not to instantly bake up a hundred pounds of hard tack for use in our home shelters.

But you can see why "survival crackers" were the food of choice when the government was looking into what to store for citizens for use in public fallout shelters. They stored well, contained plenty of calories and were a high-carbo-hydrate food that would not require lots of water and energy to digest.

Hard Tack

Hard tack is made of flour and water, sometimes with lard added, and after being baked several times at a low heat and thoroughly dried out, it will last quite a long time, which is why it was so suitable for armies on the move or ships on long voyages. It was high in carbohydrates, meaning it was a good food for quick energy, and it was relatively high in calories.

When hard tack was served to the crew on ships or a soldier in the field, it often acted as a plate for other foods. Once a man had finished off the food on the slab of hard tack, he could then begin slowly eating the slab. Often, the slab was so hard he had to soak it in his coffee or another liquid before it would soften up enough to eat.

As you can see, it is also the kind of food that can cause a dental emergency at a time when you may not be able to find or get to a dentist. This is why we don't encourage you to make a really hard version of hard tack.

Many "prepper" web sites tout the wonders of hard tack, and nobody can argue the virtues of this venerable food, but unless you have a dentist in your shelter, an event requiring you to stay in your home fallout shelter not a good time to venture into the really hard and thoroughly dried out hard tack.

Making Your Own Crackers

Many excellent cracker recipes exist, and we will give you a couple of our favorites. The major thing to look at when choosing a recipe is to avoid vegetable oils, butter or shortening which can go rancid over a long period. Remember, those old recipes used real lard for a reason, and that was because saturated fat does not go rancid.

We suggest using a good cracker recipe for a plain flour-water-salt cracker and bake it according to instructions. Test it then try baking it a second time at low heat. See how hard it is to your teeth. Then pack some in one of those round Christmas tins and put it up in your cabinet to see how long the crackers retain their freshness. You may be surprised at how long they last, unless you live in a very humid area.

Even then, a lot depends upon how you pack your crackers for long-term storage. If you pack them properly in tins or boxes, place them in Mylar bags with oxygen absorbers and desiccants then seal the Mylar bags with a hot iron, your stored crackers should last as long as the Mountain House version packed in #10 cans. For a video on this, see:

http://www.chemicalbiological.net/food%20prepar ation_july%203.wmv

Recipe 1

Norwegian Flatbread

(This bread has been made in Norway for centuries, with large quantities being prepared during the summer and stored for use during the winter months.)

3 cups rye flour

2 cups whole-wheat flour

½ teaspoon salt

About 1½ cups tepid water

Sift the flours and salt into a bowl. With a wooden spoon, stir in sufficient tepid water to form a fairly soft dough. Turn out the dough onto a floured surface and knead well for about 15 minutes, or until a little of the dough can be rolled into a cylinder and bent in half without cracking. Place the dough in a buttered bowl, cover the bowl with a damp cloth towel and let the dough rest for at least two hours.

Divide the dough into eight equal pieces and roll each piece into a round approximately 10 inches in diameter. Prick each round all over with a fork. Heat a lightly oiled griddle or a large skillet and, when it is very hot, place on of the rounds on it. Cook the bread until brown spots begin to form on the underside, then turn the bread over and brown thee other side. Reduce the heat and continue turning the bread until it becomes crisp. Repeat this process with all of the rounds. If preferred, the rounds may be place on baking sheets, pricked well and baked in a preheated 425° F oven for about 20 minutes until they are slightly brown and crisp.

For a variation, barley flour may be substituted for the rye flour. Almost any mix of flours can be used.

Makes 8 rounds.

Recipe 2

Orford's Water Biscuits

2 cups flour

½ teaspoon salt

1 teaspoon baking powder

4 tablespoons lard, cut into pieces

3 to 4 tablespoons cold water

Coarse salt

Sift the flour with the salt and baking powder, and rub in the lard. Moisten the mixture with enough water to make a firm dough. On a lightly floured board, roll the dough 1/8 inch thick, prick it all over with a fork and stamp out 3- or 4-inch rounds. Sprinkle the rounds with coarse salt and bake them in a preheated 350° F oven for 10 to 15 minutes, until the edges are pale golden in color.

Makes about 20 crackers.

Recipe 3

Hard Tack

4 to 5 cups flour

2 cups water

1 tablespoon salt

In mixing bowl, combine flour and salt. Add water slowly and mix until you have a fairly dry dough.

Roll out dough to about ½ inch thick and shape into a rectangle. Cut dough into 3 x 3 inch squares. Prick both sides thoroughly with a fork. Bake on un-greased baking sheet at 375° F for 30 minutes per side.

Store in airtight container. You can let them dry further if desired before you store them, but we do not recommend this degree of hardness for the sake of your teeth. If you happen to know someone in your group has dental problems, do not even bother with hard tack.

Makes 12 to 15 biscuits.

How Many Calories Per Cracker?

If you decide to make and store your own survival crackers, once you have decided upon a recipe, be sure and calculate the approximate amount of calories per cracker so that you can be sure of storing enough crackers to provide the amount of calories you decide upon.

Do this by adding up the calories in each ingredient to get the total amount of calories in the entire recipe then divide it by the number of crackers the recipe yields.

Bear in mind that using whole grains will result in the cracker not lasting as long as a cracker made of refined white flour. This is because a whole grain flour contains some of the oil inherent in the grain, and that oil will eventually go rancid.

This is what happened to the government's stored cans of crackers. After about 7 years of storage, some of the cans were tested and found to be rancid, most likely due to the whole wheat (graham) flour the crackers were made of.

How Many Crackers Should You Store?

If you decide to put a certain amount of survival crackers in storage in case you need your shelter at a time when there is no electricity and cooking is not an option, how can you decide how many you need?

If you decide to go with already-prepared long-term storage crackers, various companies make them. The two best known are by Mountain House and AlpineAire, two companies that make foods packed for long-term storage.

There is also a "Saloon Pilot Cracker" made by the Diamond Baking Company in Hawaii, which is an excellent cracker, but if you decide to go with these, you will likely need to repack them in metal cans to prevent rodents from chewing into them, as they come in paper cartons like regular saltine crackers.

If you really like these crackers and can afford to order plenty of them, consider storing them in Mylar bags with desiccants and oxygen absorbers also. These crackers also contain a few more calories than other brands of pilot crackers, about 65 per cracker rather than the usual 60, and you may want to take this into consideration

Each can of Mountain House Pilot Crackers or AlpineAire Royal Kreem Crackers has about 66 crackers. Some may be broken due to the way they are packed. Mountain House says each can contains 67 crackers, with 3 carded trays with 20 crackers each packed in carefully, and the spaces around them filled with 7 to 15 loose crackers. In other words, the number of crackers per can may vary. For purposes of calculation, we will assume 66 crackers per can.

Decide how many calories per day you want to allow each person in your shelter. The government assumes 750, but you may decide upon 900 or 1,000, or an even higher number. Remember, the shelter diet is merely a "keep-body-and-soul-together" diet, designed *for people following a low-physical-activity, low-water, temporary protocol for about two weeks.*

Design your shelter diet such that a one-week stay would give a fair amount of calories, and a two-week stay a lower amount. When you first go into your shelter, you will not know how long you may have to remain in the shelter. As your stay progresses, you may realize that more fallout is arriving and you'll have to extend your stay. This means you want to have made sufficient allowances so that you can change your menus mid-stay and make your food and water storage go farther. If you should have to do this, be sure and explain this to the shelterees.

Suppose that there will be five people in your home shelter, two parents and three children.

Here is an example of the calculations you would need to make if you want to be sure of having sufficient crackers on hand ... in case you are unable to cook or have few other foods on hand.

Figuring Up Your Cracker Count

To allow 900 calories per person per day:

900 calories/day = 15 crackers per day per person

5 people x 14 days (2 weeks) = 70 days of shelter stay

70 days x 15 crackers per day = 1,050 crackers needed

1050 crackers needed/66 crackers per can = 16 cans needed

You would need to store about 16 cans of Mountain House or AlpineAire crackers to meet the needs of 5 people for a 2-week shelter stay.

To allow 720 calories per person per day:

720 calories/day = 12 crackers per day per person

5 people x 14 days = 70 days' supply needed

70 days x 12 crackers per day = 840 crackers needed

840 crackers/66 crackers per can = 13 cans needed

You would need about 13 cans of crackers to meet the needs of 5 people for a 2-week shelter stay.

If you realize that your shelter stay may be longer, the moment you come to that realization is the time to begin cutting back on the daily allotment so you can extend the food supply to cover the extra days' stay.

It Never Hurts To Have A Few Extra Cans!

You can also raise your daily calorie counts by storing plenty of hard candies and jams made with real sugar. The real sugar, if properly stored, will last for years. You can put up a number of jars of jam or jelly and forget about them until you need them.

When the shelter stay is over, be sure you have prepared for the "aftertimes." This, you won't share as you did your shelter food. Have it buried out of sight. Then, you are ready for the realities of burned-out cities, a downed electric grid, and the other realities that may await you once you have survived the fallout from the nuclear war.

13: *What Is A 'Worsening Crisis?'*

How are you supposed to know when to start checking on your shelter supplies and preparing the shelter for occupation?

It's possible that you won't know, because somebody just up and does something stupid out of the clear blue sky.

Usually, however, something happens between a couple of nations that seems trivial at first, until it begins to escalate into an exchange of threats and insults, and each country starts prepping as if for war.

One such crisis was the Cuban Missile Crisis in October of 1962, when Russia placed some ballistic missiles on the communist island of Cuba in retaliation for American placement of ballistic missiles in Turkey and Italy.

The whole thing started in the summer of 1962 when Cuba's Fidel Castro negotiated with Nikita Krushchev of Russia to obtain some ballistic missiles to deter future events like the 1961 Bay of Pigs Invasion. Spy plane photos proved the existence of the missiles to the U.S., which then established a military blockade around Cuba so that no more missiles could arrive. The U.S. also demanded the missiles already in Cuba be removed.

For 13 days, from October 16 through October 28, 1962, matters seemed to hover on a razor's edge, as neither side was willing to back down. Every day the news reported statements by each side that seemed to imply that the next step was nuclear war.

Finally, after long negotiations, the Soviet Union agreed to remove the missiles, and the U.S. agreed to remove its missiles from Italy and Turkey and to never invade Cuba again without provocation. Those of you who remember the Cuban Missile Crisis probably remember a tense couple of weeks, during which each day seemed to bring a nuclear attack that much closer.

With the greatly enhanced communications abilities and spy satellites, information in this area is available almost instantly, but a true *Worsening Crisis* will develop in much the same way as the Cuban Missile Crisis.

How To Detect A Worsening Crisis

First, you will read rumors, stories in news publications that are either considered by the mainstream media to be "fake news" or headlines in mainstream media that disappear suddenly because the story was "unverifiable."

Then you will begin to read actual news stories about the growing problems with a certain country and its leadership.

The next thing you know, the White House will announce that we have a problem with the country, and we are either going to work it out or do something about it.

This marks the beginning of the crisis, which seems to worsen by the day as the countries begin trading boasts and insults and rattling their sabers in true belligerent fashion.

When you recognize a situation that you believe is developing into a worsening crisis, begin to watch the country in question. If possible, read their news publications.

If, for instance, Russia is the country causing the "problem," watch for Russia to begin moving their people and critical machinery into blast shelters. Each country has certain actions it intends to take if it believes nuclear war is imminent.

Once the crisis escalates to an exchange of conventional weapons, it may be only a matter of days or hours before one of the countries decides it wants to win at all costs and deploys nuclear weapons.

The Worsening Crisis could begin as a standoff between two smaller, non-nuclear countries, but each small country is backed by a larger nuclear country, which soon join the fray in defense of their ally.

If you identify what you believe is a rapidly worsening crisis, and you live in a large city that is a known target for a nuclear attack, or you live near important military installations or important factories for military equipment, you might want to consider taking a leave of absence or a vacation.

You need to be aware that your home shelter, unless it is underground or in a basement, is no protection against an actual nuclear blast, with all its thermal radiation and knock-down power. Your shelter is only intended to protect you from fallout radiation. This is why, if you live very close to an area that is considered a prime nuclear-strike target, you may want to consider moving or taking a trip during a major worsening crisis. *Blast shelters* are far more expensive than fallout shelters, which is why few of these were ever built in America.

What To Do In A Worsening-Crisis Situation

When a worsening crisis occurs that seems to worsen with every passing day, this is a wake-up call to you. You will want to take certain actions to enhance your chances for survival if a nuclear attack should occur while you are away from your home and your shelter.

By taking care to dress correctly and carry certain items with you at all times, you will be as prepared as possible.

Wear White!

One simple thing you can do is to wear white or pastel clothing whenever you leave your home during a time of worsening tensions. If a nuclear attack should occur, white tends to reflect the thermal, or heat radiation, whereas dark clothing tends to absorb the heat. We learned this from the atomic bomb explosions in Japan. People wearing light clothing with dark patterns imprinted on the material actually received burns to their skin that mimicked those dark patterns.

Later experiments with atomic bomb explosions in Nevada further illustrated this principle. Houses painted white smoked but did not catch fire, while houses painted darker colors burst into flames.

Carry These Items On You

A pocket respirator: Carry in your purse or pocket one of the small respirators you can order from Advanced Mart known as the Readi Mask. It combines an eye shield with a mask that filters out 99% of particulates. If you were caught in the aftermath of an explosion with the air filled with radioactive particulates, this mask seals to your face and keeps your eyes and your lungs clear of contamination. It folds into a small package you can easily carry on you.

You can buy these at:

http://store.advancedmart.com

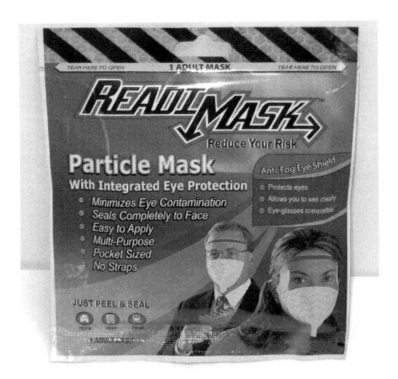

The Readi Mask

A Particulate respirator: Have in your vehicle a Moldex 2730 N100 particulate respirator, inside a plastic Ziploc bag. If you are caught in your car, or are able to get to your car, this can replace your pocket respirator if it has become clogged with particles. Have another in your office drawer in case you are caught in your office.

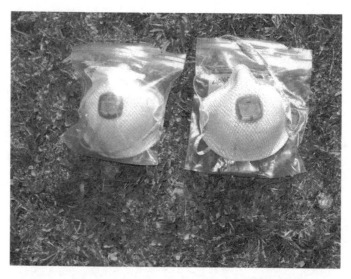

Keep A Respirator In Your Desk & Vehicle

A plastic raincoat: Many stores feature plastic raincoats or ponchos with hoods in small plastic holders that will easily fit in a purse or pocket. If an incident occurs and you must go outside, put this on over your clothes to protect yourself from radioactive particles settling in your hair and on your clothing. This is especially important if you favor a hairstyle that would trap and hold particulates, such as braids, dreadlocks, or curls, or you use a hair pomade that would serve to trap and hold particulates.

A NukAlert key fob or pocket radiological meter: If at all possible, you should have some means of detecting radiation on you. Your Civil Defense meters are usually too big to easily carry when you are going back and forth to work. The NukAlert key fob is especially good in this regard, as it will chirrup when it detects radiation and is always on duty. These are available at **http://www.ki4u.com**.

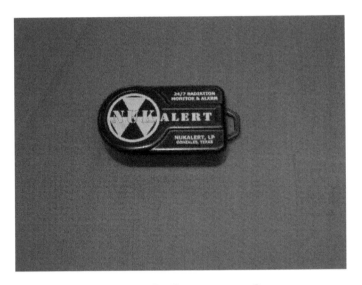

The NukAlert Key Fob

A Dosimeter: During a Worsening Crisis, make sure you have on you at all times a CD V-742 pen dosimeter, fully charged. If you should be caught out, it may be valuable to you to know exactly how much radiation you have absorbed before you were able to get to shelter.

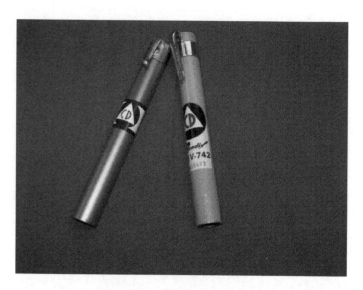

CD V-742 Dosimeters

A small radio: If you work at a desk or eat lunch at your desk, you can buy a cheap little AM/FM radio and keep it in your desk drawer. When the crisis seems to be worsening by the hour, keep a check on the hourly news reports still given by Fox News or ABC News. If a nuclear bomb explodes anywhere in the United States, the sooner you know about it, the better.

Know What To Do

Remember, the physics of nuclear science has not changed in the past 40 or 50 years. All the reporters, knowing nothing about the matter, write about the "old turn, duck, and cover" instruction as if it was outmoded—as if something better has come along to take its place.

Nothing has taken its place. If you are out on the street on some errand and you see a brilliant flash, brighter than the sun and of an unearthly hue, immediately turn away. Do not look at it to see what it is. Assume the worst and dive for the ground immediately. If there is a curb or culvert or ditch nearby you can dive into, do so. Or take cover alongside a brick building or wall.

When you take cover in this manner, you stand a better chance of missing a lot of the blast and thermal waves coming your way, with all the debris and glass traveling with them.

Stay down until you judge that the danger is past from the immediate blast effects.

Your next danger will be from immediate fallout, those heavier particles that rain down immediately after the blast. Immediately put on your rain scarf and your pocket respirator and make your way to better cover inside a building, or even your car if it will run.

If your car is running and the electronics haven't been fried by the nuclear EMP, or if you drive a pre-1976 model without electronic ignition, you are in luck. You can head toward your home as far as you can get before the streets or freeways become clogged with stopped vehicles and force you to a halt.

If the fallout is heavy, you may need to get inside a building and wait until the heaviest downfall is over. The less exposure you have to immediate fallout, the better.

Once the atmosphere has cleared of most of the particulates, if your radiation meter indicates low readings, it may be safe for you to head for your home shelter. If the readings are high, you might wish to remain inside a building for a day or two until the radiation has a chance to decay.

129

As soon as possible, head to your home and your home fallout shelter.

There, your own careful preparations await you, and you are ready to ride out the period of time it will take all the fallout to decay. This will depend upon the proximity of the blast, how many blasts there are and the direction of the wind.

But once you're home, inside your home shelter, you are as safe as you can make yourself, and dependent upon no one for your personal safety or that of your family.

The laws of physics in all of this *have not changed.*

The actions you should take for survival have not changed either.

Do not be fooled by the "publish-or-perish" science of today which would have you believe that there is nothing you can do to save your own life and your family's.

Many years ago, researchers discovered that many serious burns and injuries in the atom-bombed Japanese cities of Hiroshima and Nagasaki could have been avoided if only the Japanese had known to "turn, duck and cover."

A few simple actions like the "old" turn, duck and cover, can still save your life!

Radiation Safety Limits

0 to 50 Roentgens	Considered Safe
0 to 200 Roentgens	Level I Radiation Sickness
200 to 450 Roentgens	Level II Radiation Sickness
450 to 600 Roentgens	Level III Radiation Sickness

Level I Radiation Sickness: Less than 5% deaths. From 5% to 30% of exposed people may develop acute symptoms of nausea and vomiting within 4 hours of exposure. A temporary reduction in blood platelets and white blood cells may occur.

Level II Radiation Sickness: Less than 50% deaths. From 60% to 75% of exposed people may develop acute symptoms nausea and vomiting within 4 hours of exposure. Severe blood changes, hemorrhage, and hair loss.

Level III Radiation Sickness: Greater than 50% deaths. One-hundred percent of exposed persons will develop acute symptoms of nausea and vomiting within 4 hours of exposure. Severe blood damage, hemorrhage, and hair loss, with up to 80% deaths in less than 2 months.

Generally speaking, the higher the dose of radiation received, the faster the onset of symptoms. At Chernobyl, the doses received were so high, firefighters noted a metallic taste in their mouths and a severe headache before they developed nausea and vomiting that forced them to stop their work.

In a couple of "criticality accidents" in experimental labora-tories (cases where an unexpected event caused a brief, high release of radiation), the radiation-blasted person immedi-ately felt as if he was burning and tingling all over and devel-oped almost instant nausea and profuse vomiting.

As has been said by many experts in the field, if the radiation dose is high enough for you to actually feel, it's also high enough to kill you.

Here is a small chart you can copy onto a 3 x 5 card and keep with your meter. In an actual incident, stress tends to de-grade your memory, and radiation is something you do not want to take chances with.

1 Roentgen = 1 Rem = 1 Rad = 0.01 Gray = 0.01 Sievert

0 – 30,000 CPM = Safe

Greater than 30,000 CPM = Questionable

1R = 1Rem = 1Rad = 0.01Gy (Gray) = 0.01 Sv (Sievert)
0 - 30,000 CPM = Safe
Greater Than (>) 30,000 CPM = Questionable

Resources

Actions For Survival: Saving Lives in the Immediate Hours After Release of Radioactive or Other Toxic Agents, 2011, Allen Brodsky

A Field Guide to Radiation, 2012, Wayne Biddle

Chernobyl 01:23:40: The Incredible True Story of The World's Worst Nuclear Disaster, 2016: Andrew Leatherbarrow

Cookies & Crackers, The Good Cook: Techniques & Recipes Series, Editors of Time-Life Books, 1982

Dictionary of Gastronomy, 1970, Andre L. Simon & Robin Howe

Family Shelter Designs, Department of Defense, Office of Civil Defense, January, 1962.

Instruction and Maintenance Manual for Dosimeter Ratemeter CD V-736, Dosimeter CD V-746, Dosimeter Ratemeter Charger CD V-756: The Bendix Corporation

Instruction and Maintenance Manual: Radiological Survey Meter (CD V-715), 1962: The Victoreen Instrument Company

Introduction to Radiological Monitoring: A Programmed Home Study Course, HS-3, 1972: Staff College, Defense Civil Preparedness Agency

LaRousse Gastronomique, Edited by Jenifer Lang

Operating and Maintenance Instructions: Radiological Dosimeter Jordan Model 750-5, Jordan Electronics, a Division of the Victoreen Instrument Company

Overcrowding Potential, **Robert A. Krupka, June, 1964. Office of Civil Defense, Department of the Army.**

Radiation Defense Textbook, 1963, Department of Defense, Office of Civil Defense

Radiation Safety In Shelters, 1983: Federal Emergency Management Agency

Standard Supplies For The Fallout Shelter (Annex E)

About the Author

Dr. Charles S. Brocato is a scientist and author who has written extensively in the fields of surviving chemical and biological warfare and nuclear warfare. He is also a longtime martial arts instructor, weapons/firearms instructor, and a map & compass instructor, specializing in *how to stay alive*. He is also a nutritionist and a counselor in the field of nutrition, and an award-winning French chef.

He holds degrees in biology and mathematics with advanced studies in the areas of biochemistry, microbiology, and advanced mathematics, and a D.D. (Doctor of Divinity). He has also completed numerous studies in areas of Emergency Management with FEMA and MetEd.

Two of his previous books are available on Amazon in print form: *The Two-Fold Chastisement: Visions of the Coming Earth Changes*, and *Chemical/Biological WarFare: How You Can Survive*.

This book is the second in a series of Radiation titles he has planned in the area of surviving a nuclear attack or incident.

The first book of this series, *How to Choose A Civil Defense Radiological Instrument: Geiger Counters & Dosimeters*, is available on Amazon.com in print and e-book forms.

Write him at: **csbauthor@chemicalbiological.net**

Dr. "B"s Motto:
"You Can Never Have Too many Meters!"

Made in the USA
Middletown, DE
12 October 2017